Aquaponics

_____

Aquaponics

Combining Aquaculture and Hydroponics

Edited by

Pierre Foucard and Aurélien Tocqueville

CABI Quæ éditions

CABI is a trading name of CAB International

CABI
Nosworthy Way
Wallingford
Oxfordshire OX10 8DE
UK

CABI
200 Portland Street
Boston
MA 02114
USA

Tel: +44 (0)1491 832111
E-mail: info@cabi.org
Website: www.cabi.org

Tel: +1 (617)682-9015
E-mail: cabi-nao@cabi.org

Originally published in French under the title *Aquaponie: Associer aquaculture et production végétale* edited by Pierre Foucard and Aurélien Tocqueville © Éditions Quæ, 2019.

A catalogue record for this book is available from the British Library, London, UK.

ISBN-13: 9781836991427 (hardback)
        9781836991434 (ePDF)
        9781836991441 (ePub)

DOI: 10.1079/9781836991441.0000

Commissioning Editor: Rebecca Stubbs
Editorial Assistant: Emma McCann
Production Editor: James Bishop

Translator: DeepL
Typeset by Straive, Pondicherry, India
Printed in the USA

Contents

———————

Introduction

---

1. The development of aquaculture worldwide and in France

Over the last twenty years or so, world fish stocks have remained stable at around 90 million tonnes, including fish for human consumption and industrial fishing. Given the current state of wild fish stocks and the ecological impact of overfishing, no significant increase in fishing quotas is likely. The future supply of aquatic products to the world market to meet the growing demand for human consumption therefore relies on the development of aquaculture, which is now the fastest-growing animal food production activity in the world, with over 6% growth per year between 1986 and 2016 (FAO, 2018) (Figure i-1). In 2015, aquaculture accounted for around 53% of the supply of fish for human consumption, compared with 14% in 1985. According to FAO forecasts, this boom will continue to meet the growing demand for fish protein, which is increasing as the world's population grows (FAO, 2018). Asia is the most dynamic continent in the development of aquaculture, with 42% of total aquaculture production coming from this economic activity (75% for China), compared with just 18% for Western Europe (FAO, 2018).

Despite growing demand and recognised expertise, the French fish farming industry is not experiencing any significant growth, mainly due to major regulatory obstacles. As a result, France is now over 85% dependent on imported aquaculture products.

2. Questions raised by aquaculture activity

Aquaculture regularly suffers from a poor reputation in French public opinion, and is mainly criticised for the way it feeds fish and the pollution it generates. According to a report drawn up by Hélène Tanguy as part of a mission for the French Ministry of Agriculture and Fisheries on the development of aquaculture (30 October 2008), "exchanges providing scientific information on the change in feed composition in favour of plant substances, or guarantees of satisfactory control of downstream rivers or the seabed have not succeeded in changing intellectual attitudes". However, over the last ten years or so, there have been many encouraging announcements about aquaculture: "Aquaculture is a sector where there are not enough investment projects. The development of this economic sector must be a priority for France, which must enrich its strategy for supplying freshwater fish products" (Michel Barnier, 2007); "Fish are the future of mankind" (Jean-Paul Besset, MEP, 2012); "There must be a clear political will to develop sustainable and competitive aquaculture, in order to

Millions of tonnes

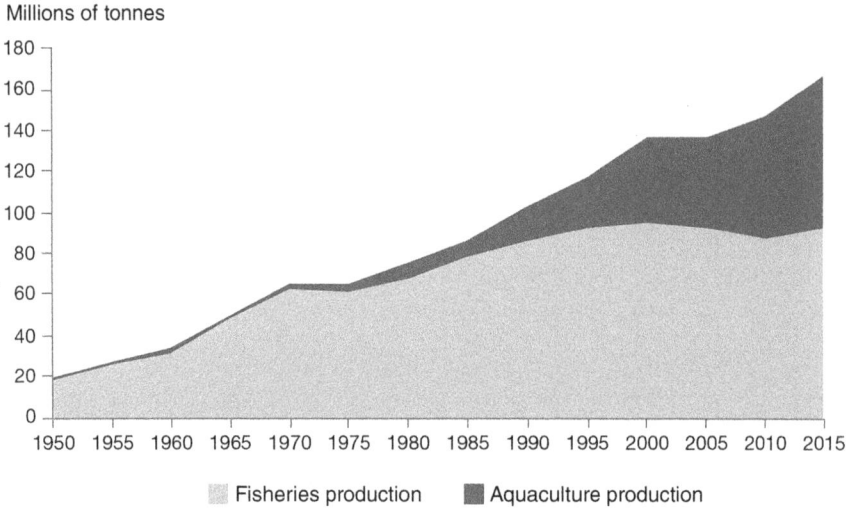

Figure i-1. Evolution of aquaculture production and capture fisheries activities, 1950-2016. (FAO, 2018)

face up to competition from third countries" (Alain Cadec, 2012)[1] ; "French aquaculture is an activity capable of satisfying the three pillars of sustainable development, it creates long-term activity on the coasts and in rural areas, generates skilled jobs that cannot be relocated and is part of genuine regional projects (Martinie-Cousty *et al.*, 2017). Despite this, fish farming production in France has remained stable, with very few new installations over the last twenty years.

If aquaculture is to be given a new lease of life and become more popular in French society, it must face up to the challenge of integrating more satisfactorily into the environment it is helping to change, and must bring out new paradigms in favour of ecologically intensive production. "Over the next ten years, total production from aquaculture and fisheries will surpass that from beef, pigs and poultry" (Árni M. Mathiesen, FAO Assistant Director-General, 2013), provided that "better ecosystem management is encouraged". The development of aquaculture requires greater consideration to be given to the efficient use of the food and environmental resources available to us. The idea is therefore to complement existing production systems where they are deficient: this means above all developing more sustainable fish feed and better health management, as well as developing more efficient technologies based on saving water resources and recycling effluent.

3. The emergence of new paradigms: recirculating circuits and integrated multi-trophic aquaculture systems

Recirculating" aquaculture systems have been developing in Europe for a number of years, particularly in Denmark. They aim to recycle and reuse the water leaving fish farms in order to control discharges into the environment and limit dependence on this resource. Other research is focusing on integrated multi-trophic aquaculture (IMTA) systems that make the most of water bodies rich in nitrogen and phosphorus that come out of seawater or freshwater fish farms. To achieve this, plants, algae or even molluscs are co-produced with a fish farming compartment, so as to reclaim what was previously considered a simple waste product. Aquaponics can therefore be seen as an example of an AIMT system in freshwater and seawater. It combines a recirculating fish farming circuit with the cultivation of plants for human consumption to create a complementary economic activity, which distinguishes it from conventional phytopurification (planted marshes, lagoons, etc.).

3.1. Aquaponics, a growing method inspired by ancestral practices

Around the year 1200, the Aztec civilisation cultivated gardens in a lake environment called "*chinampas*", consisting of artificial islands, generally rectangular, whose surface emerged about 1 m from the surface of the water. This word of Nahuatl origin (an indigenous Mexican language) literally means "place of the reed fence". It is sometimes translated as "floating garden". These structures are held together by a network of rushes, reeds and foliage, covered on the surface by mud from the bottom of the lakes, rich in decomposing organic debris, all arranged in successive layers (Turcios, 2014). It is difficult to say to what extent the fish in these waters helped to fertilise the water, but the concept of 'soilless' plant cultivation was born at this time. Today, this type of cultivation has all but disappeared. Some plots have been preserved in the Xochimilco district of Mexico City and are now on UNESCO's World Heritage List (Figure i-2).

Long before that, around 1700 years ago, rice-fish culture systems appeared on the Asian continent in mainland China (Renkui *et al.*, 1995). This was an integrated system for producing rice and fish (traditionally carp, eel or tilapia) (Figure i-3). This practice has lasted through the ages, after centuries of existence in various Asian countries, while remaining very much in the minority due to the work involved in modifying the structure of existing rice fields for this purpose (Edwards, 2015). Today, rice-fish farming is still practised in Bangladesh, China and Vietnam, notably with giant freshwater prawns (*Macrobrachium rosenbergii*), Chinese crabs (*Eriocheir sinensis*) and Louisiana crayfish (*Procambarus clarkii*) (Edwards, 2015).

3.2. Modernising the concept of integrated fish and plant farming

The modern concept of aquaponics and the term 'aquaponics' itself emerged in the 1970s and 1980s following research carried out by the New Alchemy Institute in North Carolina, which focused on the development of intensive organic farming techniques. It was demonstrated that water from fish farms was an interesting source of nutrients for hydroponically-grown crops (Todd, 1980; Zweig, 1986). This institute no longer exists, but its publications are still used

Figure i-2. Radish growing on chinampa near Lake Xochimilco (David Arqueas)

Figure i-3. Combined rice and tilapia aquaculture in a rice field, Yogyakarta, Indonesia (Kembangraps)

as references today[2]. Other North American research institutes followed suit a few years later (Sneed *et al.*, 1975; Naegel, 1977; Lewis *et al.*, 1978).

Inspired by the success of the New Alchemy Institute and the theme of aquaponics, Mark McMurtry of the University of North Carolina pursued this line of work by developing a system of vegetable cultivation combined with tilapia farming in the 1990s, introducing the issues of water conservation, intensive fish protein production and reduced operating costs (Mc Murtry, 1997). At the same time, Dr James Rakocy of the University of the Virgin Islands (UVI) developed a semi-commercial-scale system that is now the benchmark for sizing the fish and plant compartments and has been operating continuously for many years (Rakocy *et al.*, 2006). This 'UVI system' is best known for having transferred a reproducible model, which has been adapted by various commercial operators.

Modern technology offers prospects for the development and diversification of the fish and vegetable industries. Today, it is a serious avenue of study for complementary and alternative methods of aquaculture and vegetable production. Without seeking to replace what already exists, it is more often than not designed to make a place for itself on non-agricultural land (non-fertile land, industrial wasteland, abandoned market gardening greenhouses, urban and peri-urban areas, etc.) and to position itself on a short circuit market.

3.3. The development of aquaponics around the world

Considered one of the "ten technologies that could change our lives" by the European Union Parliament (Woensel and Archer, 2014), research into aquaponics still suffers from a lack of insight into the economic reality of this large-scale activity. However, it has been the subject of considerable dynamism in the United States, Canada and Australia for some twenty years, and in Europe since the 2010s. The keen interest in this production method is even reflected in the many initiatives by private individuals to replace their traditional vegetable gardens with small-scale aquaponics production systems, encouraged by the many resources available on the internet: specialist blogs and *Youtube* channels (*The Aquaponics Journal*; *BackYard Aquaponics*; *Bright Agrotech*) to name but a few.

Few references exist on the number of commercial-scale aquaponics businesses, and it is difficult to estimate the fish and vegetable production carried out in this way on a global scale. There are still few commercial systems in the world. However, we can mention a few: Cultures Aquaponics inc. and Hydronov in Canada; Urban Organics, Superior Fresh, Florida Urban Organics in the United States; GrowUp Urban-Farms and BioAquafarm in England; ECF Farmsystems in Germany; BIGH in Belgium, and De l'eau à la bouche, Ferme aquaponique de l'Abbaye, Nutreets in France.

Companies specialising in the advice, design and sizing of aquaponic systems have also sprung up alongside these commercial farms, such as Aquaponic Solutions (Australia) and Nelson and Pade (United States), and have become benchmarks in their field.

According to a 2014 study based on a survey of 809 "aquaponiculturists" around the world (the only existing survey of this type), 80% of aquaponics producers are in the United States, making it the leading country in the field, 8% in Australia and 2% in Canada (Love *et al.*, 2014), with the remaining 10% in the rest of the world. This study is not exhaustive and is already out of date, particularly for Europe.

Until the years 2012-2013, Europe was "lagging behind" what had been done on the American, Australian and Asian continents in terms of research and development. Resolution 2013/2100 (INI) was adopted by the European Parliament on 11 March 2014 (Mc Intyre, 2014), as part of reflections on the future of the horticultural sector in Europe and strategies for growth. This resolution mentions that "aquaponics systems hold potential for local and sustainable food production and can contribute, by combining freshwater fish farming and vegetable growing in a closed system, to reducing resource consumption compared to conventional systems". Europe caught up rather well between 2014 and 2018, thanks to several research projects that emerged at the same time:

– The INAPRO (Innovative Aquaponics for Professional Applications) research project, led by the Institute for Freshwater Ecology in Leibniz, began in 2014. This project brings together 18 partners across 8 countries to develop a large-scale aquaponics system that is economically and ecologically viable, and innovative compared with systems currently in existence around the world;

– The EU Aquaponics Hub European network, active from 2014 to 2018, was set up by the COST (European Cooperation in Science and Technology) inter-governmental programme. It has brought together various research players: IGFF (Institute of Global Food and Farming) in Denmark, Svinna-verkfraedi Ltd in Iceland, Nibio (National Institute of Biomedical Innovation; formerly known as "Bioforsk") in Norway, the Ponika company in Slovenia, the Eureka Farming company in Italy, The FishGlassHouse at the AUF (Faculty of Agricultural and Environmental Sciences) in Rostock in Germany, the Zurich University of Applied Sciences (ZHAW) and the Tropenhaus company in Switzerland, the PAFF Box at the University of Gembloux Agro-Bio Tech, Nerbreen and the University of Las Palmas de Gran Canaria in Spain. A collaborative European map has been set up by the COST network, on which any company, individual, association or university can add and locate its aquaponics system. It can be consulted on the COST network website[3]. This map is not exhaustive for France, as many private projects emerged between 2013 and 2018, some of which have already been implemented. Most of them are referenced on another collaborative online map set up as part of the APIVA® project, specifically for France. This can be consulted on the APIVA® project website[4];

– the APIVA® project (Aquaponics, plant innovation and aquaculture) led by ITAVI, in partnership with INRA, ASTREDHOR, CIRAD and EPLEFPA de Lozère, financed by CASDAR (Special allocation account for agricultural and rural development) as part of the 2013 call for innovation and partnership projects, ran in France from 2014 to 2017. Several feedback sessions were held during this period, bringing together a growing number of stakeholders in the field of aquaponics in France: project developers, consultancies, researchers, government departments, chambers of agriculture and the simply curious. The aim of the project was to set up medium-scale experimental pilots (from 60 to 200 m² of production area), to characterise the compartments of these aquaponics systems and to study the flows using modelling approaches, with the aim of establishing elements for dimensioning and technical and economic efficiency, but also to study the quality of the products and the purification yield of this type of system, and finally to disseminate knowledge to the fish, horticulture and market gardening industries. Figures i-4, i-5 and i-6 illustrate the three experimental pilots set up in the partner structures of the APIVA® project. Since 2018, the APIVA® project has continued with new lines of research thanks to a project supported by the FEAMP (European Maritime Affairs and Fisheries Fund).

Figure i-4A and B. RATHO experimental pilot at Brindas. (Pierre Foucard, ITAVI)

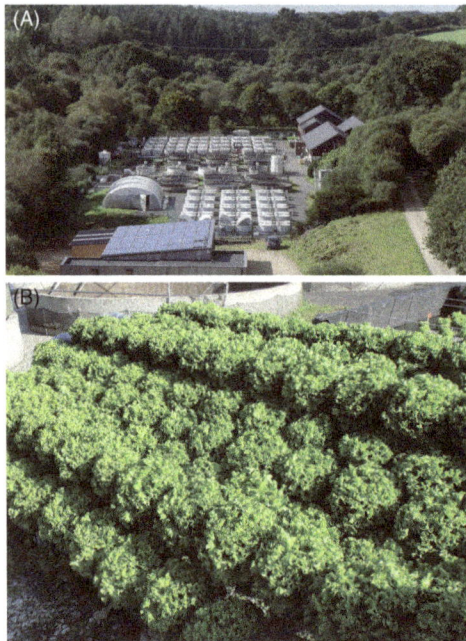

Figure i-5A and B. Inra-Peima experimental pilot at Sizun. (Victor Dumas, Inra)

Figure i-6. Experimental pilot at EPLEFPA de Lozère in La Canourgue. (Catherine Lejolivet, EPLEFPA de Lozère)

Notes

[1] For M. Barnier (2007), J.P. Besset (2012) and A. Cadec (2012), see the website of the Centre d'études pour le développement d'une pisciculture autonome.

[2] For more information: http://www.thegreencenter.net (consulted on 25/01/2019).

[3] See https://euaquaponicshub.com/eu-aquaponics-map/ (consulted on 28/01/2019).

[4] See https://projetapiva.wordpress.com/le-projet-apiva-objectifs-et-partenaires/ (consulted on 15/03/2019).

1

Aquaponics: concept, approaches and uses

Aquaponics combines two distinct and complementary production methods: soilless plant cultivation (based on hydroponics) and aquaculture (based on recirculating systems). Each of these production methods has its advantages and limitations, and it is necessary to understand how they work in order to combine them in the best possible way. The central idea is that the waste from one compartment becomes a resource for the second, and that the combination of the two means that less input (water and fertiliser) is required. Based on this principle, the design of the system can vary greatly depending on whether you are starting from scratch or from an existing structure, and on the final objective (commercial, educational, self-production), which will determine the scale and intensity of production. The aim of this first chapter is to take stock of these different factors.

4. Components of aquaponics: hydroponics and aquaculture

4.1. Hydroponics, or soil-less cultivation

4.1.1. Principle of hydroponics

Hydroponics refers to vegetable or horticultural production carried out in a "soilless" environment, usually under glass (Figure 1-1). The plants grow outside the field, on a solid, neutral and chemically inert substrate (rock wool, coconut fibre, peat, pine bark, nutrient films, etc.), where irrigation takes place on a periodic or continuous cycle to keep the roots in a moist environment and provide a constant supply of nutrients.

Plants are fed by the circulation or percolation of a nutrient solution that provides the water, dissolved oxygen and mineral elements essential for plant growth, all under controlled and regulated pH and conductivity conditions.

Originally developed in Holland in the 1940s and 1950s, the soil-less cultivation process is now spreading throughout the world. Generally applied in greenhouses, it allows many plants to be grown outside their area of origin, increasing yields while ensuring consistent product quality, and improving working conditions. It does, however, involve significant investment and profitability can be difficult to achieve.

4.1.2. Overview of the soil-less sector in France

In 2012, the total area under ornamental horticulture in France was 17,957 hectares, including 1,903 hectares under greenhouses and tunnels and 2,156 hectares of soilless production platforms, i.e. 12% of the total area (France Agrimer, 2013). In the market garden sector, soilless greenhouse cultivation is currently the preferred method for producing several species belonging to the fruiting vegetable category, such as cucumbers and tomatoes (for the fresh market). These two crops currently account for more

©2026 CAB International. *Aquaponics* (eds Pierre Foucard and Aurélien Tocqueville)
DOI: 10.1079/9781836991441.0001

Figure 1-1. Growing tomatoes hydroponically in a greenhouse, above ground on coconut bread and drip irrigation (Carlos yo)

than 70% of cultivated areas. Strawberries are also being grown under glass (Agreste, 2013). In 2014, around 35,000 tonnes of strawberries were produced off-grid in France, representing 20% of the surface area and over 50% of production volumes (Réussir Fruits et Légumes, 2014).[5]

4.1.3. Advantages of hydroponics

This production technique provides a high level of control over inputs, the quantity and quality of the water used (mineral concentration, pH, conductivity), and environmental parameters (temperature, air humidity, light), while taking advantage of the greenhouse effect.

What's more, hydroponics eliminates the problems sometimes associated with soil (pathogens, salinity, pH, soil structure), makes it possible to grow plants on non-arable land in urban areas (gardens, roofs, balconies, etc.) and peri-urban areas (industrial wasteland), eliminates the conventional practices of tilling the soil and weeding, and simplifies cultivation techniques, while increasing yields.

Intensive hydroponic crops can achieve yield optimisation of 20 to 25% compared with soil-based crops (Somerville *et al.*, 2014), in particular thanks to the absence of root competition, which means that crop densities are often higher than in conventional agriculture. Lastly, the cultivation itineraries are much simpler in soilless systems than in conventional systems, and enable efficient and unrestrictive harvesting, particularly with the help of cultivation structures placed at ground level.

4.1.4. Limits of the hydroponic technique

Traditional soilless plant production technologies use nutrient solutions rich in mineral fertilisers, often in excess, to provide the plants with all the elements they need for growth. The system is regularly drained when nutrient imbalances occur in the growing water: this drainage represents 20 to 50% of the volume of nutrient solution supplied, and up to 80% in cut flower production (Bron, 2012). Soil-less cultivation with lost drainage has been on the decline since the early 2000s; the most advanced soil-less cultivation systems tend to use technologies that recycle the nutrient solution (Boulard, 1999). In this case, surplus nutrients not used by the plants are a source of raw material for the manufacture of new nutrient solutions. The mineral composition of the drained solution is not constant, as it depends on the relative rate of uptake of each ion by the plants, so it is necessary to take this into account when making compensations. Despite this, the recycling of solutions is not systematically practised because of the investment required and the technical skills involved. In addition, excessive recycling of fertiliser solutions can lead to an accumulation of certain minerals or organic acids (such as benzoic acid) and plant exudates (allelochemical compounds) that are toxic to plants (Hosseinzadeh, 2017). It is therefore impossible to close the system down 100%, and it is essential to empty the tanks periodically to facilitate the return to the desired chemical balance and limit the health risks (Bron, 2012).

Another limitation of hydroponics is its dependence on the use of fossil fuels, whether because of the massive use of plastics or because of the way in which simple nitrogenous chemical fertilisers or other complex mineral salts are synthesised (Haber process for nitrogenous forms, mining of phosphate and potash and subsequent chemical treatments to obtain solubilised forms).

4.1.5. Recirculating aquaculture system (RAS)

4.1.5.1. OVERVIEW OF THE AQUACULTURE SECTOR. In 2016, global aquaculture production stood at 80.04 million tonnes, including 54.09 million tonnes of freshwater and marine fish. Global fish supply reached a record 20 kg per capita, thanks to strong growth in aquaculture. China accounts for over 60% of global aquaculture production. Aquaculture now supplies just over half of all fish for human consumption (FAO, 2018). Fish farming systems in so-called "open" or "pond" environments are predominant in the world for freshwater aquaculture.

France is one of Europe's leading producers of rainbow trout (32,200 tonnes in 2015, France Agrimer) and marine fish fry (5,000 tonnes of sea bass, sea bream, turbot and meagre), more than half of which is exported. There are also 6,000 tonnes of carp and other freshwater fish produced in ponds. Trout is the main product, produced using traditional open-farming systems.

4.1.5.2. PRINCIPLE OF RECIRCULATING AQUACULTURE. Recirculating" aquaculture systems are conceptually opposed to conventional aquaculture systems in circuits or "open" environments.

In conventional freshwater fish farming systems (trout farming, for example), the water is constantly renewed in the ponds to ensure good water quality for the farmed fish, good oxygenation, and elimination of suspended solids and certain dissolved molecules (ammonia, nitrate, orthophosphates) through a "dilution" approach. It should be noted, however, that this water is "used" and not "consumed", as the principle is to divert part of the flow from a river (and/or a borehole and/or a spring) to the rearing tanks and then return it to the watercourses after use. In this type of system, managing solid (suspended solids) and dissolved (nitrogen and phosphorus) waste is problematic because of the high flow rates. Similarly, other open systems such as cage farming at sea or continental ponds are dependent on the conditions of the external environment. Increasingly stringent European regulations (often applied more harshly in France) are placing a heavy burden on these economic activities and hampering their development because of issues relating to effluent discharges, ecological continuity, and water usage rights.

In contrast to conventional 'open' systems, 'recirculated' systems have been developed over the last thirty years, and their field of use is gradually expanding from the hatchery to the grow-out phase, for all types of farming (Timmons and Ebeling, 2007). They enable a large proportion of the water used for farming to be recycled following various purification and regulation treatment stages using mechanical and biological filters to eliminate suspended matter, detoxify nitrogenous ammoniacal nutrients and control dissolved gases. Sterilisation of the water using ultraviolet radiation or, in some cases, ozone treatment is generally carried out to limit the risk of introducing exogenous pathogenic organisms (Klinger and Naylor, 2012; Labbé et al., 2014). These recirculated systems also offer the possibility of optimising rearing conditions (T°, O_2, sanitation, etc.) by freeing them from the conditions of the external environment. Figure 1-2 illustrates the general principle of a "RAS" system.

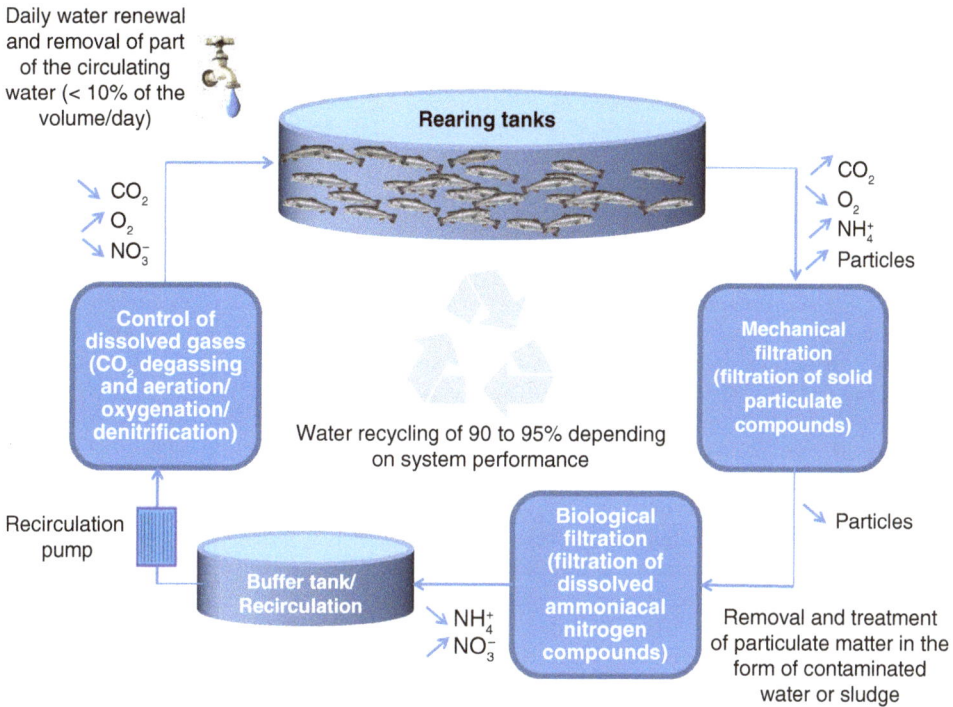

Daily water renewal and removal of part of the circulating water (< 10% of the volume/day)

Rearing tanks

CO_2
O_2
NO_3^-

CO_2
O_2
NH_4^+
Particles

Control of dissolved gases (CO_2 degassing and aeration/ oxygenation/ denitrification)

Mechanical filtration (filtration of solid particulate compounds)

Water recycling of 90 to 95% depending on system performance

Recirculation pump

Buffer tank/ Recirculation

Particles

Biological filtration (filtration of dissolved ammoniacal nitrogen compounds)

NH_4^+
NO_3^-

Removal and treatment of particulate matter in the form of contaminated water or sludge

Figure 1-2. Operating principle of a recirculating aquaculture system (RAS) (Pierre Foucard, ITAVI)

Despite the interest in this practice, and the growing number of countries gradually adopting this technology, its contribution to overall aquaculture production is still low in Europe (Labbé *et al.*, 2014) and worldwide compared with conventional aquaculture.

4.1.5.3. ADVANTAGES OF RECIRCULATING AQUACULTURE. Recirculating systems have the advantage of allowing better control of rearing water parameters than conventional open systems. They also minimise the rate of water renewal in the system, which generally varies between 1% and 10% of the rearing volume per day (Blidariu *et al.*, 2011; Timmons and Ebeling, 2007) depending on the biomass loaded and the amount of feed distributed in the system.

They reduce the volume of effluent discharged by the fish farm by 90-99% compared with a conventional system, making it easier and more efficient to treat the effluent. As a result, the environmental impact of fish farming is greatly reduced. This technique makes it possible

to farm fish on sites where water resources are limited, while ensuring partial decontamination of effluents (suspended solids and ammoniacal nitrogen).

In addition, recirculated circuits allow better control of incoming water quality due to the low flow rates used. This makes it possible to use water disinfection techniques based on ultraviolet radiation or ozone.

4.1.5.4. LIMITS OF RECIRCULATING AQUACULTURE. The investments involved in this type of system can be quite substantial (Schneider *et al.*, 2006), while energy costs are increased and represent around 10% of operating costs, compared with 5% in a traditional open circuit system, which would require intensified production of at least 100 tonnes/year to be economically viable (Rey Valette, 2014; Lennard, 2018), preferably with a species offering high added value. However, it is possible to achieve small-scale, profitable systems by producing fry and juveniles for grow-out in conventional fish farming.

Despite zootechnical progress in the field of fish feed (optimisation of feed digestibility and fish conversion index), only 20 to 50% of the nitrogen and 15 to 65% of the phosphorus contained in the feed are used by the fish in the digestion process, which means that the remaining portions are not assimilated but rejected by the animals (Schneider *et al.*, 2005) and constitute waste.

Recirculated aquaculture involves lower volumes of effluent discharge than conventional systems, but for the same species and size of fish produced, the overall quantities of nitrogen and phosphorus produced are identical. Although the effluent is smaller in quantity, it is also more concentrated, making it more efficient to treat. Today, denitrification technology can transform up to 95% of residual nitric nitrogen (nitrates) into atmospheric nitrogen, but this process is very complex to control and there are still doubts about the gaseous nitrogen form that results: Nitrogen N_2 is harmless and in fact makes up 78% of the chemical composition of the earth's atmosphere, which is not the case for nitric oxide N_2O, which is a major greenhouse gas; there is currently little data on the gaseous forms (particularly the N_2/N_2O ratio) released by this type of nitrate treatment system in the context of aquaculture. Phosphorus emissions are very difficult to treat.

4.2. Aquaponics, combining hydroponics and recirculating aquaculture

4.2.1. Principle of aquaponics (Figure 1-3)

Aquaponics can be defined as a combination of an aquaculture compartment (usually with fish reared in a recirculating system), a plant compartment (above-ground) and a bacterial compartment (biological filter), all in a closed or virtually closed circuit.

Aquaponics enables the bacterial and plant compartments to make the most of aquaculture effluent, thereby reducing dependence on water supplies and totally or partially reducing the use of synthetic fertilisers for plant production. Dissolved waste from aquaculture is made up of solid particles, ammonium, phosphorus, potassium and other macro- and micro-elements. These elements are sources of nutrients that can be assimilated by plants after a preliminary stage of microbial degradation of ammonia compounds into nitrates by nitrifying bacteria, and

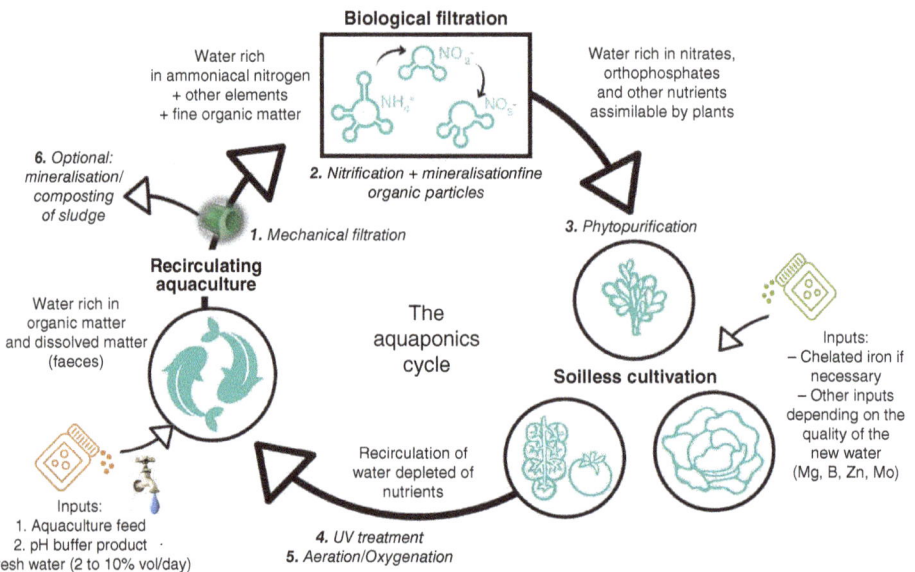

Figure 1-3. Operating principle of an aquaponics system. (Pierre Foucard, ITAVI)

elimination of particulate matter by mechanical filtration. These steps enable the water to be recycled and made safe for fish farming, plant cultivation and the development of colonies of nitrifying bacteria.

4.2.2. Advantages of aquaponics (Figure 1-4)

Aquaponics is attracting a great deal of interest because of its many advantages:

–   better management of water resources: dependence on water resources for crop production is much lower than for conventional open-field farming, with a reduction in consumption of the order of 90 to 99% (Somerville *et al.*, 2014);

–   recovery of fish farming waste: whereas hydroponics requires mineral inputs (of chemical and mineral origin) to be added to the nutrient solution to meet the requirements of the plants, the aim of an aquaponics system is to do away with these inputs and replace them with the metabolic waste

from associated fish farming. Unlike open or recirculated aquaculture systems, aquaponics has the advantage of recovering effluents laden with dissolved compounds from aquaculture production, making them bioavailable as nutrients for soilless plant production (Rakocy *et al.*, 2006; Diver, 2006; Klinger, 2012);

–   higher plant yields and easier cultivation: soilless techniques produce higher yields than conventional cultivation (Rakocy *et al.*, 2006; Savidov, 2005; Saufie *et al.*, 2015; Delaide *et al.*, 2016), with a 20-25% increase in plant yield, and even 2 to 5 times higher productivity for some crops (Somerville *et al.*, 2014). Several studies show that there is no significant difference in yield between hydroponics and aquaponics (Savidov *et al.*, 2005; Graber and Junge, 2009; Pantanella *et al.*, 2010; Thorarinsdottir *et al.*, 2015). Others claim that aquaponics supplemented with hydroponic fertilisers can have higher plant yields than conventional hydroponics, highlighting the role of the microbial community in the

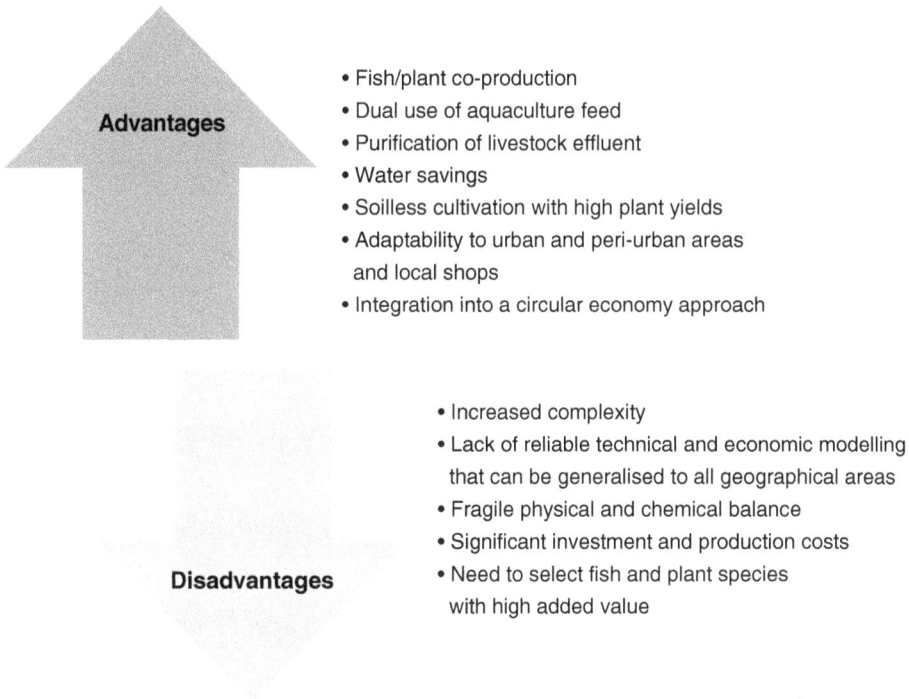

Advantages

• Fish/plant co-production
• Dual use of aquaculture feed
• Purification of livestock effluent
• Water savings
• Soilless cultivation with high plant yields
• Adaptability to urban and peri-urban areas
  and local shops
• Integration into a circular economy approach

Disadvantages

• Increased complexity
• Lack of reliable technical and economic modelling
  that can be generalised to all geographical areas
• Fragile physical and chemical balance
• Significant investment and production costs
• Need to select fish and plant species
  with high added value

Figure 1-4. Main advantages and disadvantages of aquaponics. (Pierre Foucard, ITAVI)

effectiveness of aquaponics (Delaide *et al.*, 2016). Finally, recent studies show that fish farm effluents can hinder the development of certain plant pathogenic fungi, notably *Pythium ultimum* and *Fusarium oxysporum* (Fujiwara *et al.*, 2013; Gravel *et al.*, 2015). This resistance could be due to the presence of a stable and ecologically balanced environment, with a wide diversity of microorganisms, some of which are thought to be antagonistic to the pathogens that affect plant roots and some of which could help plants to grow while forming a barrier against pathogens;

– cost sharing: aquaponics allows land, infrastructure and equipment to be shared between two distinct economic activities. It also makes it possible to smooth out the production costs associated with aquaculture feed, energy, water and labour over two production compartments operating in series. The fish feed, which is also used as fertiliser for the plants, is used twice and is therefore potentially more profitable than with a fish compartment alone, provided that the plant compartment is sufficiently sized for the economic reality. It should be noted that this sharing of costs can not only take place within a single company, but is even more relevant when two professionals join forces to combine fish farming with market garden production under glass, making two totally independent activities but operating in synergy, in a win-win partnership;

– use of areas that are of little or no use to agriculture: since aquaponics is based on saving and reusing water, it offers the possibility of producing fish and plants in regions where the soil is poor and access to water resources is limited, or even in arid and semi-arid regions (Diver, 2006; Somerville *et al.*, 2014). In addition, an aquaponic greenhouse can be set up in urban and peri-urban areas close to places of consumption, which encourages the development of a local economy based on short circuits, thereby limiting the costs and CO_2 emissions associated with transport (Diver, 2006). This new activity also has the potential to develop an educational and social dimension.

4.2.3. Limits of aquaponics (Figure 1-4)

4.2.3.1. AQUAPONICS IS STILL IN ITS INFANCY AND FACES A NUMBER OF CHALLENGES:

– the complexity of coupling aquaculture and market gardening; merging two production systems doubles the potential for technical problems. Aquaponics requires technical skills in a number of areas: aquaculture, market gardening and/or horticulture, water chemistry as well as electricity and hydraulics. It therefore requires a skilled and trained workforce (Nemethy, 2016) to deal with the various problems that can arise. It is important to understand that many factors influence the dynamics of the system, making it difficult to transpose from one site to another without modelling and sizing work upstream. The water quality of the resource, the climate, the fish species and its development stage, the plant species and its phenological stage, the feeding strategy applied, the composition of the feed, the thermal and energy aspects (filtration, lighting of plants, thermoregulation of livestock and crops, etc.) are all parameters likely to influence the performance of animal and plant production;

– the creation of links between two types of production... a double-edged sword. It is important to point out that the use of plant protection products - for the plant compartment - and antibiotics - for the aquaculture compartment - is not recommended without an in-depth study of the potential risks of toxicity to fish and bioaccumulation by plants respectively (Zhang *et al.*, 2017). The fact that pesticides cannot be used means that biological control solutions that are compatible with animal life will have to be found. Furthermore, antibiotics could have an inhibiting effect on the biofiltration process carried out by nitrifying bacteria, which is essential for the recirculating system to function properly (Fredricks, 2015). The fact remains that not using antibiotics in this type of system entails major economic risks in the event of a disease outbreak in the stock, given the lack of truly effective alternatives, and requires increased monitoring of the fish farm and systematic prophylactic measures;

– a lack of perspective on economic profitability. Aquaponics has emerged through the ideology of sustainable food production, not through market demand. The capital investment required to design these systems is generally significant, and the investment in time through the learning process also needs to be taken into account. With the exception of very specific situations (arid/semi-arid regions, regions with a favourable climate throughout the year, etc.), the profitability of an aquaponics system remains questionable, and the dimensions that would enable it to be achieved are still difficult to grasp given the current state of knowledge. Energy costs (linked to the operation of pumps, oxygenation systems, filtration or water disinfection) are often higher than those already existing in hydroponics, while also including them (light and heating) depending on the situation. Many systems currently in existence are not economically viable and could not survive without various forms of subsidy or activities to complement production. Those that have succeeded, however, and which can be considered as commercial installations, have used highly effective niche marketing techniques through retail sales and/or diversification into ecotourism, equipment supply, training, advice and expertise, while aiming for a sufficiently large scale of production. Today, aquaponics has to overcome technical, economic and social obstacles, and the start-up of such a sector requires the support of research and training structures. Labelling to enable products from aquaponics to be promoted would probably be a *sine qua non* for making this production method profitable. The development of a detailed financing plan, backed up by a market study - on the most appropriate plant and aquaculture products in the context of a given region, targeting products with high added value - should therefore not be overlooked in the context of commercial installations. Available data on existing commercial systems in European climates suggests that a minimum scale of 1,000 m^2 is required to achieve profitability. Unfortunately, it is still very difficult to obtain more details on the technical and

economic parameters, given the often confidential nature of this information. It should be noted, however, that plants are often more profitable than fish per unit area, and as such, the "plant" compartment of a commercial aquaponics system can make a small-scale recirculating system economically viable, which would not otherwise be possible (Lennard, 2018).

5. Different approaches to aquaponics

Aquaponics therefore has a wide range of interests, with the main aim of meeting environmental challenges and ensuring that new production technologies are accepted. If it is to have a real future - over and above its usefulness from an educational point of view - the various approaches that can be taken to this technique need to be adapted to the specific context of the various economic players who could one day be involved in its development. These players are not only fish farmers, horticulturists and market gardeners, but also very often project developers who are totally unfamiliar with these sectors.

It is therefore possible to adopt the point of view of the fish farmer, who is primarily looking to treat his effluents, the point of view of the market gardener, who is primarily looking to use less synthetic fertiliser, or that of the producer - fish farmer or market gardener - who is looking to optimise both productions on an economic scale.

5.1. Piscicultural approach

From the point of view of fish farming, aquaponics is used to treat solid and dissolved waste from the water of existing fish farming installations, using systems that combine filtration elements that are preferably simple and inexpensive, since the majority of production remains fish farming. The APIVA® experimental pilot at the Inra-Peima station at Sizun in Brittany is based on this approach: a hydroponic culture system has been connected to the outlet of a pre-existing recirculated fish farming system: trout/vegetables/leaves, 65 m^3 of rearing and 84 m^2 of crops (Figure 1-5). For the fish farmer,

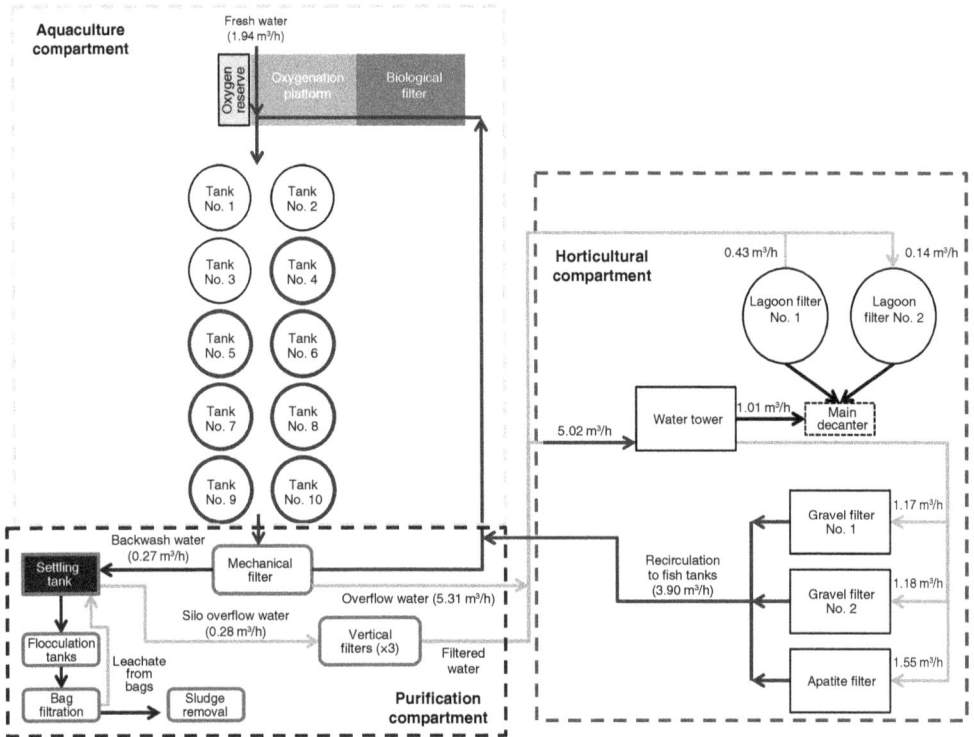

Figure 1-5. Schematic diagram of the aquaponics system installed at the Inra-Peima experimental station in 2011. (Inra-Peima, 2018)

the aquaponics system thus represents an effective means of reducing the level of nitrogen and phosphorus molecules present in the water before discharge into the watercourse.

The aim of this approach is not to achieve a complete balance between fish and plant production, or to make the plant compartment a major economic entity. The technical choices for the market-garden section are therefore geared towards a *low-cost* design. Some of the water can easily be reintegrated into the aquaculture system by means of various effluent purification processes (mechanical, biological, phytodepuration), resulting in significant savings for the fish farming industry, which is extremely dependent on the flow of watercourses (problems of reserved flows/withdrawals), which themselves fluctuate greatly depending on the season. The challenge is therefore mainly regulatory and environmental, in order to comply with the requirements of the Water Framework Directive (2000/60/EC), which aims to reduce water consumption and the discharge of eutrophying

molecules into aquatic environments. This approach has also already been considered in trials of ferti-irrigation of field crops using water from fish farms, without leading to commercial developments.

5.2. Plant approach

From the market gardener's point of view, the farm's main activity is plant growing. A fish farming compartment is associated with a pre-existing soilless vegetable crop. Two independent recirculated systems are therefore combined with the aim of optimising water use and reducing or even completely abandoning the use of N-P-K nutrient solutions, which have a significant environmental impact, from their manufacture (chemical or mining industry) to their release into the environment (during washing and emptying of the systems) (Afsharipoor *et al.*, 2010). Here again, the aim of this approach is not necessarily to completely balance the fish and plant compartments, but to feed the plants with an 'organo-mineral' nutrient solution, either as the main input or as a partial supplement to a conventional hydroponic nutrient solution.

It is possible to imagine this configuration as a partnership between two producers, each with their own trade - fish farmer and market gardener - and wishing to join forces as part of an "exchange of good practices": the fish farmer can thus entrust the management of part of his waste to a market gardener, while the latter can reduce his consumption of inputs and water.

5.3. Mixed approach

Following this approach, the aquaponics system will be designed to balance the fish and plant compartments, which requires technical knowledge and perfect mastery of two very different yet complementary activities, both of which will then have to be optimised in terms of sizing and considered as two economic activities of equivalent importance to the farmer. The experimental APIVA® pilots at RATHO and EPLEFPA de Lozère -

developed as part of the APIVA® project - are based on this principle, and aim to balance fish farm waste with plant requirements as far as possible and to do without N-P-K fertiliser altogether (Figure 1-6). This approach is the one most often targeted by project developers outside the fish farming and market gardening sectors, who are usually starting from scratch.

This design represents a genuine integration of two food production systems, with the aim of minimising water consumption and producing plants on a large scale with the sole aid of fish effluent. The two activities are designed to be mutually profitable. Rigorous management must be applied to handle the effluents without risking excessive accumulation of certain nutrients, while covering all the nutritional requirements of the plants. This requires very precise modelling and technical and economic sizing of the system prior to the implementation of such a project, unlike the approaches presented above which accept a degree of 'flexibility' (incomplete effluent purification for the fish farming approach, partial coverage of plant needs for the plant approach).

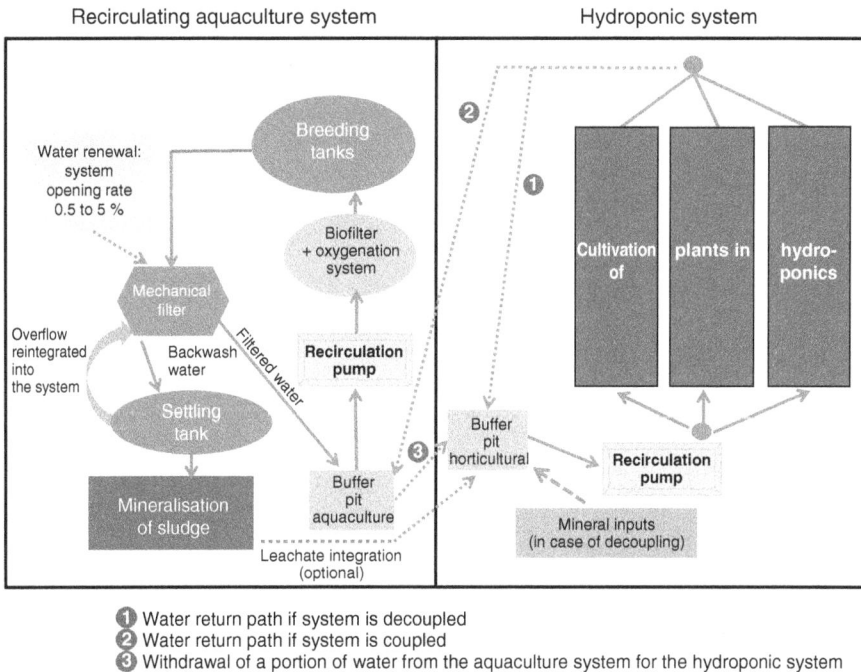

Figure 1-6. Schematic diagram of aquaponics systems installed at the RATHO experimental station and EPLEFPA de Lozère in 2014 and 2015. (ITAVI)

6. Different scales, different objectives

The term "aquaponics" is used to describe a wide range of systems with very different objectives. Just as aquariology and fish farming are two radically distinct activities that cannot be transposed, so a home vegetable garden and a commercial-scale market gardening business do not cover the same issues. It therefore seems useful to define scales of aquaponic production.

6.1. Production scales

Depending on the scale of production and the intended purpose (commercial food production, leisure, concept demonstration and education, subsistence feeding, etc.), the level of technical skill will not be the same.

6.1.1. Domestic or very small-scale aquaponics (5 to 20 m²)

This scale of system is suitable for individuals wishing to produce part of their own food, or simply for recreational use. These home systems generally comprise a small livestock tank combined with a small area of plant cultivation on gravel or NFT (technical nutrient film) for scales of between 5 and 20 m². The FAO has drawn up a reference document for the design of this type of system, available online at.[2]

6.1.2. "Social" or "demonstration" aquaponics, on a small or medium scale (50 to 200 m²)

These rather "low-tech" systems are not profitable from the point of view of production alone, and often rely on ancillary activities to ensure that they operate economically: training, site visits, setting up ecotourism structures for educational purposes, etc. These structures are generally presented as "demonstrators" and are usually located in urban areas, on the roofs of buildings or on local authority land in built-up areas (gardens, parks, schools, etc.). These structures are generally presented as "demonstrators" and are usually located in urban areas, on the roofs of buildings or on local authority land in built-up areas (gardens, parks, schools, etc.).

6.1.3. "Semi-commercial" aquaponics on a medium scale (200 to 1,000 m²)

These systems have a commercial production objective and incorporate a level of technical expertise that enables them to produce up to a few tonnes of fish per year for production areas of varying sizes. The economic models for these systems are often fragile and highly dependent on niche markets, while the products are usually sold directly to consumers as part of a short distribution chain. As in the previous category, income is generally supplemented to a significant degree by activities ancillary to production.

6.1.4. Large-scale "commercial" aquaponics (1,000 to 4,000 m² or more)

These highly sophisticated systems generally target large scales and incorporate several large rearing and culture tanks, filtration (mechanical and biological) and oxygenation systems, alarms and automation. They can be installed in suburban areas on industrial wasteland, or in rural areas on sites already identified for aquaculture and on which it is possible to combine vegetable crops, or on market garden production sites on which it is possible to install recirculating aquaculture circuits. The scale of production makes it easier to achieve viable economic models, provided that market knowledge is mastered and products are promoted through labelling.

6.2. Level of production intensity

Farming and cultivation systems can be categorised according to their level of intensity, which is generally linked to their scale of production, but also to the density of farming (for the fish farming part, expressed in kg/m³) and the density of cultivation (for the vegetable part, expressed in plants/m²). The level of intensity will determine the technical nature of the equipment needed to operate them. The larger the scale of the system, and the higher the density of livestock and crops, the more technical the equipment will need to be. There are three levels of intensity: extensive systems, semi-intensive systems and intensive systems.

6.2.1. Extensive systems

Extensive systems generally have no specific equipment dedicated to biological and mechanical filtration, although clay balls and other culture media may be sufficient in adequate quantities to filter out suspended matter, which will then enrich the medium after microbial degradation; fish stocking density is generally very low (≤ 10 kg/m^3 as an indication) but very dependent on the design method and the species reared.

6.2.2. Semi-intensive or middle-tech systems

They operate with an inexpensive passive mechanical filtration system (such as a settling cone and/or filtration foam) or even a bead or sand filter, and must include a specific compartment dedicated to biological filtration. The stocking density rarely exceeds 20 kg of fish per cubic metre of water due to the limited efficiency of the mechanical filtration system: a higher density would require a higher water renewal rate to avoid an excessive increase in the amount of organic matter circulating in the system, which would limit the system's performance in terms of water consumption.

6.2.3. Intensive or high-tech systems

They consist of a finely dimensioned biological filtration system and a high-performance mechanical filter (of the drum filter type) coupled or not with passive filtration. They also include technologies for aerating the water by stirring (when the fish stocking density is low, ≤ 50 kg/m^3) and/or oxygenating the water by adding liquid oxygen (when the fish stocking density is high, ≥ 50 kg/m^3 up to a tolerable biological limit specific to each fish species). The mechanical filtration system is highly efficient, allowing a high closure rate for the system. Once concentrated, this sludge can be used as manure on agricultural land, or as an input for methanisation, or even recovered by lombrifiltration or biodigestion (Goddek *et al.*, 2016b) with a view to reintegrating some of these nutrients into the aquaponic system.

7. Different design strategies: coupled or decoupled systems

There are different ways of designing aquaponic systems, the coupled system and the decoupled system. The classic (coupled) aquaponics system consists of a fish farming circuit shown in dark grey in figure 1-7, which is directly connected to the hydroponics unit shown in light grey. Water circulates constantly between the fish farming circuit and the plant growing area. The decoupled aquaponics system consists of a fish circuit connected to the hydroponics unit via a one-way valve (Figure 1-7). A recovery tank independent of the fish recovery tank allows water to be recirculated in the plant compartment only, making it possible to have two recirculation loops - fish and hydroponics - independent of each other, where the fish water is simply supplied to the hydroponics unit as required, leaving open the possibility of returning the water to the fish compartment according to the conditions defined by the producer.

7.1. Coupled systems

Coupled" aquaponics systems are those designed in the early days of aquaponics, notably at the New Alchemy Institute, North Carolina State University and the University of the Virgin Islands (UVI).

The disadvantage of this coupling strategy is the total dependence between the two aquaculture and hydroponic compartments: the water leaves the fish tanks, passes through a mechanical filtration system before reaching the biofiltration compartment, then continues its journey through the plant system before returning to the fish compartment. This dependence between aquaculture and plant production makes it difficult to maintain optimum living conditions for all the living elements in the system (plants, bacteria, fish) and requires very precise dimensioning. It implies a constant compromise in the face of a very fragile and evolving physico-chemical balance. In the event of a health problem affecting fish or plants, solutions are limited, as treatment products for plants and fish are not always compatible.

Even 'organic' plant protection products can have an impact on fish health. Similarly, anti-parasite or antibiotic products for fish are incompatible with the cultivation of plants for human consumption, in the absence of information on the capacity of plants to 'bio-accumulate' certain compounds. This need to make compromises and the lack of control over production

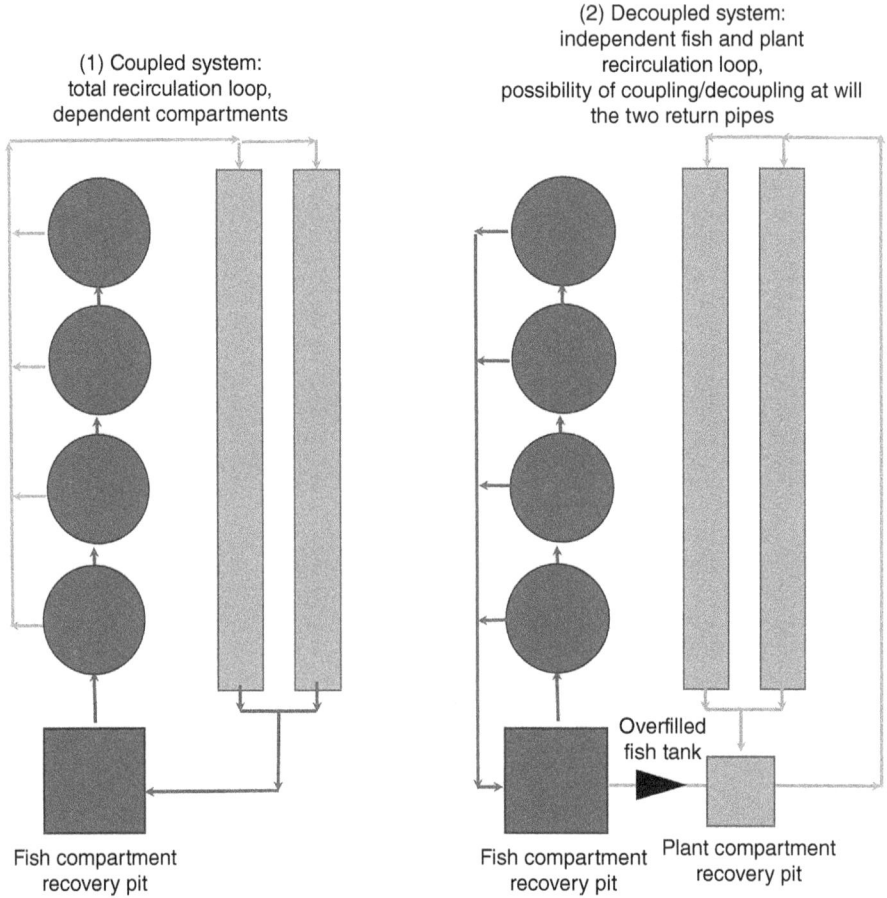

Figure 1-7. The different modalities for designing aquaponic systems: the coupled system (1) and the decoupled system (2). (Monsees *et al.*, 2017a)

are the main obstacles to the commercial application of aquaponics, limiting aquaponics to small- to medium-scale systems for the most part.

7.2. Decoupled systems

Current research efforts are aimed at developing so-called "decoupled" systems arranged in two separate and independent loops (fish and hydroponics). Water of fish origin is recirculated only within the plant unit, allowing better control of the specific requirements of the animal and plant species (Monsees *et al.*, 2017a).

The loss of water due to evapotranspiration from the plants is compensated for by channelling water from the fish farms into the hydroponic tank. These systems can be designed to be coupled/decoupled on demand; it is therefore not outlawed in this type of system to return water from the plants to the fish, depending on the production strategies employed. Decoupled systems are generally more flexible in their operating mode, and safer for commercial production, since in the event of problems with the fish farming system (prophylactic treatments, pest control treatments, mechanical, electrical or filtration problems, etc.), it is possible to make the fish farming hydraulic circuit independent of the hydroponic hydraulic circuit by decoupling the two compartments, while the fish are being treated (Yildiz *et al.*, 2017).

In a "decoupled" strategy, the water from the hydroponic unit may or may not be redirected into the rearing tanks, depending on the growing strategy (whether or not additional fertilisers are added, whether or not acidifying products are added to improve nutrient uptake by the plants). The physico-chemical conditions within the hydroponic unit can be managed separately, so that the fish are not subjected to excessively sudden changes in their living environment (water acidification, increased iron or potassium concentrations). In this configuration, the water from the fish farm should provide most of the inputs required for plant growth, with the option of adding additional fertilisers as required to promote plant growth.

Monsees *et al.* (2017a) report a 36% improvement in productivity in 'decoupled' aquaponics compared with 'coupled' aquaponics on a tomato crop, thanks to the fact that the decoupled configuration allows control of pH and nutrients, as close as possible to the plants' preferred nutrient levels. Similarly, Delaide *et al.* (2016) compared three lettuce growing methods: pure hydroponics based on mineral inputs (HP), pure aquaponics (AP) consisting simply of water from a recirculated fish farm circuit, and aquaponics supplemented with mineral inputs (CAP) so as to obtain nutrient concentrations similar to the HP method. The results are interesting in that the HP and AP modalities gave very similar yield results, while the CAP modality resulted in a 39% increase in growth rate at the end of the study. It should also be noted that the root biomass was more developed for the CAP and AP modalities than for the HP modality, which suggests that the fish farm effluents really do provide added value for the plants. It may therefore be necessary to supplement the aquaponics water with fertilisers (potassium in particular) to grow fruit plants and/or nutrient-hungry vegetables (such as aubergines) if we wish to obtain yields similar to hydroponics and avoid deficiencies in the leaves and fruit (Vergote and Vermeulen, 2012; Goddek *et al.*, 2015; Delaide *et al.*, 2016; Suhl *et al.*, 2016; Monsees *et al.*, 2017a). For less nutrient-intensive plants (leafy plants, aromatic herbs), the results obtained as part of the APIVA® project do not seem to indicate any need to supplement water with elements other than iron.

8. Aquaponics: an evolving basic concept

Aquaponics has traditionally been designed, tested and developed for use in freshwater recirculating systems. But should we limit ourselves to this configuration? On the one hand, it is entirely possible to consider aquaponics with brackish or salt water. On the other hand, the use of an open circuit, rather than a recirculated circuit, would seem to be feasible.

8.1. Aquaponics in salt water?

Research work has been carried out in France by the CTIFL (Centre technique interprofessionnel des fruits et légumes) on growing tomatoes in lightly salted water (Laurent Rosso, CTIFL, personal communication), with NaCl levels of 1 to 5 g/l. These salt levels are, by way of example, compatible with European sea bass production, which tolerates levels of 0.5 g/l to 37 g/l (Rubio, 2005) subject to gradual acclimatisation. We could therefore envisage aquaponics combining these two types of production, which would be particularly interesting for fish farming given the added value of sea bass.

Adding salt water to the growing water is a practice sometimes used in hydroponics, particularly for tomato crops. The increase in osmotic pressure induced by salinity reduces water consumption by the plants and allows a greater concentration of minerals in the plant tissues, which ultimately has positive effects on the taste, nutritional values and shelf life of the product. Similarly, the production of certain medicinal plants is enhanced by saline tension, which increases the concentration of active molecules used in the pharmaceutical industry and the nutritional quality of certain plants such as cherry tomatoes, which produce more antioxidants and vitamins C and E to protect themselves from the resulting water stress (Sgherri, 2008). The salinity level tested in this study (conducted in Australia) corresponded to a conductivity of 10 mS/cm, obtained with seawater diluted to 12%, which corresponds to a salinity of 4 g/l. This opens up prospects for diversifying plant production by co-cultivating with marine species (fish or shrimp) in salt or brackish water. Mariscal-Lagarda *et al.* (2012) tested the cultivation of tomatoes using effluent from white-legged shrimp (*Litopenaeus vannamei*) in very low salinity water - in the region of 0.8 to 1 g/l - and compared the tomato yield with that of conventional hydroponics. Tomato yields were similar between aquaponics and hydroponics.

It should also be noted that polychaete worms - for example the species *Nereis diversicolor* - can be reared in slightly brackish to salty water in a fairly simple way in a culture tank with a layer of sand at the bottom to allow the worms to form galleries, and with a diet based on organic matter (fish sludge).

This type of farming could provide a financial supplement through the sale of worms as fishing bait and/or as a food supplement for fish, given the richness of these worms in polyunsaturated fatty acids (Pajand, 2017). Other worms, such as arenicolae (*Arenicola marina*), are also of commercial interest to the pharmaceutical industry because of their 'universal' haemoglobin composition, which could be used to make 'artificial blood' compatible with human blood.

8.2. Aquaponics in an open fish farm?

It is traditionally considered that the *sine qua non* condition for producing plants in aquaponics is to operate with a recirculated fish circuit, the reasoning being that closing the system leads to an increase in conductivity, an accumulation of nitrogen and phosphate nutrients and other macro- and micro-elements, and a high contact time of the water with the plants, in short optimal physico-chemical and hydraulic conditions. The portion of water leaving recirculated systems is characterised by a high concentration of nitrogen and phosphorus and a low flow rate, while the water leaving open systems has low nitrogen and phosphorus levels (due to dilution of the water from the farm) and flows at a high rate.

However, Buzby *et al.* (2016) have succeeded in growing a wide variety of plants (lettuce, herbs, mustard, watercress, rocket, etc.) from seed on *raft* systems fed with 'conventional' fish farm effluent, i.e. in an 'open' circuit, and therefore with water that is very low in nutrients, up to 200 times less concentrated in nitrogen

and phosphorus than what can conventionally be found in water from recirculated circuits.

From this experiment, we can see that the quantity of nutrients available over a given time period is certainly greater than the concentration of nutrients at a given point in *time*: in other words, some plants can grow with water that is very low in nitrogen and phosphorus as long as the flow rate under the planted surfaces is high enough to renew the nutrients dissolved in the water in contact with the plant roots. It would be interesting to acquire more data on the nutritional quality and conservation capacity of plants grown with water that is very low in nutrients, and therefore with very low conductivity. Furthermore, the authors of this publication do not specify the average weight of the plants harvested (lettuces) or the presence or absence of signs of nutritional deficiencies on the leaves. Generally speaking, it is difficult to attest to the reality or otherwise of the technical feasibility of aquaponics in an open environment, given the paucity of bibliographical references on the subject, but this avenue deserves to be explored, given that the vast majority of French fish farming systems are in open circuits and not in recirculated circuits, which makes aquaponics adaptable to very few existing sites.

9. Conclusion

The concept of aquaponics is a very old method of cultivation, but its recent commercial application makes it a particularly complex and demanding cutting-edge technique. This production method is designed to meet the regulatory and environmental challenges facing two hitherto unrelated industries: aquaculture and soil-less market gardening/horticulture. Aquaponics is not set in stone. Whether an aquaponics system is in fresh or brackish water, in a coupled or decoupled circuit, the objectives remain the same: to turn what was once considered a waste product into a valuable input in order to limit the impact of aquaculture and plant production on the environment thanks to the bioremediation carried out by the plants and the total or partial abandonment of the use of nitrogen and phosphate fertilisers, while limiting dependence on water resources.

Table 1-1. Summary of the technical data and characteristics inherent in the different scales of production and levels of intensiveness of aquaponic systems, with regard to the technical design choices generally implemented. (Pierre Foucard, ITAVI) Fish, plants, bacteria: three technical worlds to understand

	Very small scale	Small to medium scale	Medium scale	Large scale	Very large scale
Goal	Hobby	Demonstration/education	Pre-commercial scale	Commercial scale	Commercial scale
Configuration	Coupled	Coupled	Coupled/decoupled	Uncoupled	Decoupled/additional inputs
Intensive production	Extensive	Extensive to semi-intensive	Semi-intensive to intensive	Semi-intensive to intensive	Intensive
Technical	Low-tech.	Low-tech	Low-tech to semi-high-tech	Semi-high-tech to high-tech	High-tech
Surface area allocated to market garden production	15 to 20 m² of floor space	100 to 200 m² of floor space	200 to 1000 m² of floor space	1000 to 3000 m² of floor space	3000 to 10,000 m² of floor space
Average biomass of "stock" fish	50 kg	250 to 500 kg	500 kg to 2.5 T	2.5 to 8 T	8 T to 25 T
Average fish farming density	10 kg/m³	20 kg/m³	40 kg/m³	50 kg/m³	70 kg/m³
Volume of fish farming	5 m³	15 to 25 m³	12 to 60 m³	50 to 160 m³	160 to 360 m³
System opening rate	Evaporation compensation	300 to 1000 l/kg of feed	300 to 1000 l/kg of feed	300 to 500 l/kg of feed	100 to 300 l/kg of feed
Main source of income	Self-production/none	Visits, training	Visits, training, plant production	Plant production	Fish and plant production
Mechanical filtration	Media/foam	Media/passive drainage	Decanting + mechanical filtration	Drum filter + decantation	Drum filter + decantation
Biological filtration	Culture media/foams/roots	Static biological filter/culture media	Biological filter on fluidised bed	Biological filter on fluidised bed	Biological filter on fluidised bed
Oxygen supply	Water movement, bubbling	Ventilator	Ventilator	Ventilation and/or oxygenation	Oxygenation
Water disinfection	/	UV filter	UV filter + prophylaxis	UV and/or ozone filter + prophylaxis	UV and/or ozone filter + prophylaxis

Embarking on such an undertaking implies having an overview of the different possible models of approach, with intensities and scales of production closely linked to very specific objectives (hobby, demonstration/education, commercial application) (Table 1-1).

Notes

[5] See https://www.reussir.fr/fruits-legumes/fraisereferences-le-hors-sol-fait-redecoller-la-fraise (accessed 11/02/2019).

[6] See http://www.fao.org/3/a-i4021e.pdf (consulted on 11/02/2019).Environment, Host and Pathogen: Disease Epidemiology

2

Fish, plants, bacteria: three technical worlds to understand

What species of fish and plants can be produced in aquaponics, and what are their particular needs? What technologies are needed to operate a recirculating aquaculture system? How important is water filtration and replenishment? What plant cultivation techniques are applicable to aquaponics? How can we recognise nutritional deficiencies in plants? How important is the microbial compartment in the system? These are all questions that are fundamental to any commitment to aquaponics, and to which this second chapter provides answers. The aim is to provide an overview of the different compartments that make up an aquaponics system and to provide some technical bases for understanding how they work.

10. The aquaculture compartment: zootechnics and technology

10.1. Which aquaculture species should be raised in aquaponics?

According to a statistical study carried out worldwide on the diversity of aquaponic farming practices, the fish species most commonly raised in aquaponics are tilapia (*Oreochromis mossambicus* and *Oreochromis niloticus*), catfish (*Ictalurus punctatus*), sun perch (*Lepomis gibbosus*), rainbow trout (*Oncorhynchus mykiss*) and largemouth bass (*Micropterus salmoides*) (Love *et al.*, 2014). But many other fish species have been tested in recreational or production facilities around the world: striped bass (*Morone saxatilis*), brown trout (*Salmo trutta fario*), Arctic char (*Salvenilus alpinus*), sturgeon (*Acipenser baerii*), European perch (*Perca fluviatilis*), baramundi (*Lates calcarifer*), zander (*Sander lucioperca*), bream (*Abramis brama*), Murray cod (*Maccullochella peelii*), reed carp (*Ctenopharyngodon idella*), pacu (*Colossoma macropomum*), koi and hybrid carp (*Cyprinus* spp.), scalaria (*Pterophyllum scalare*), guppy (*Poecilia reticulata*) and even certain species of prawn or crayfish (Diver, 2006; Rakocy *et al.*, 2006; Connoly and Trebic, 2010; Endut *et al.*, 2010; Tyson *et al.*, 2011).

The choice of species depends on how easy it is to supply fry, the resources and quality of the water, the climate and temperature that the farmer is able to maintain, and the regulations specific to each country in the case of fish species that are not locally represented, exotic or invasive. But in reality, it is above all market demand that will be the determining factor.

10.2. Basic zootechnics

10.2.1. The concept of fish farm density

This factor is highly dependent on the species and the technical nature of the fish farming

©2026 CAB International. *Aquaponics* (eds Pierre Foucard and Aurélien Tocqueville)
DOI: 10.1079/9781836991441.0002

system. The density of intensive fish farms typically fluctuates between 30 and 60 kg/m³. In closed systems, some farms adopt densities of over 100 kg/m³.

10.2.2. Nutrition and zootechnical performance measurements

10.2.2.1. NUTRIENT REQUIREMENTS. Farmed fish need a high-quality, balanced diet with high nutritional value if they are to grow quickly and stay healthy. They are fed rations containing all the necessary nutrients, in varying proportions depending on the species: proteins (30-50%), lipids (10-25%), carbohydrates (15-20%), phosphorus (<1.5%), as well as vitamins and minerals (Craig and Hellfrich, 2009). For fish at a high trophic level, feed is composed partly of fishmeal (between 17 and 65%) and fish oil (between 3 and 25%), marine resources with a high nutritional value (FAO, 2018) but also constituting a limited resource because they are overexploited.

The quantity and composition of the feed will vary during the growth cycle of the fish for a given species, depending on the growth stage: diets for juvenile fish are higher in protein (40-50%), while diets adapted for adult fish have a lower protein content (30% for tilapia, 40% for trout). Figure 2-1 illustrates the wide variability in the average proximal content of the various components of fish feed, while Figure 2-2 shows the ranges of protein levels (in %) recommended in the diets of different species of fish according to their physiological needs at the grow-out stage.

It is essential to manage livestock nutrition as a function of various zootechnical indicators: ration rate, growth rate and feed conversion ratio (Guillaume *et al.*, 1999).

10.2.2.2. CONSUMPTION OR CONVERSION INDEX (CI). CI is a unitless parameter that is a ratio between the quantity of feed distributed (kg) and the gain in fish biomass (kg). It is an indicator of fish feeding efficiency. In short, it corresponds to the "quantity of feed required (kg) to produce 1 kg of fish". It can be chosen by the farmer on the basis of the results usually observed and set as a target to be achieved. It varies essentially according to the temperature of the rearing water, the species, the age of the fish, the quality of the feed (digestible energy and composition) and its digestibility. The average consumption index for a farmed rainbow trout varies from 0.8 to 1 over the entire cycle. The feed conversion ratio increases as the animal grows.

10.2.2.3. SPECIFIC GROWTH RATE (SGR). It corresponds to the daily gain in mass acquired by a

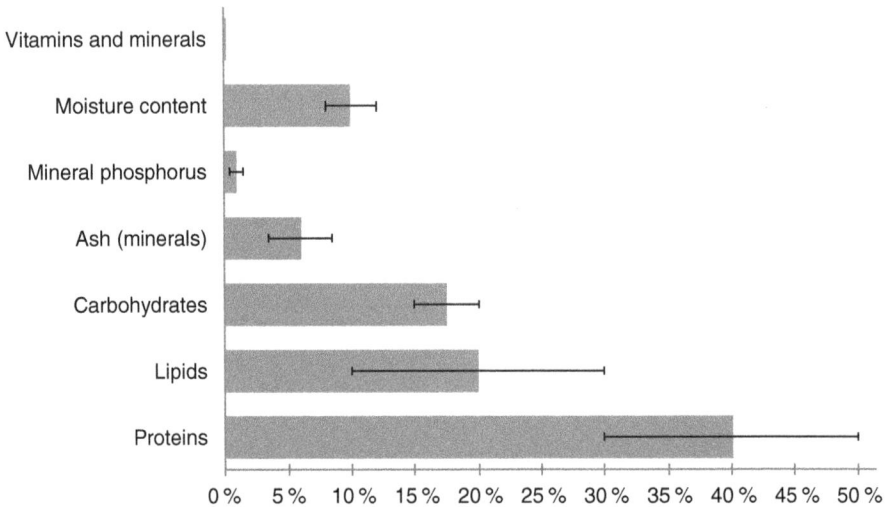

Figure 2-1. Variability in the average proximal content of the various components of fish feed. (Pierre Foucard, ITAVI, based on data from Craig and Hellfrich, 2009)

Recommended protein content ranges (%)
according to fish species (growth stages)

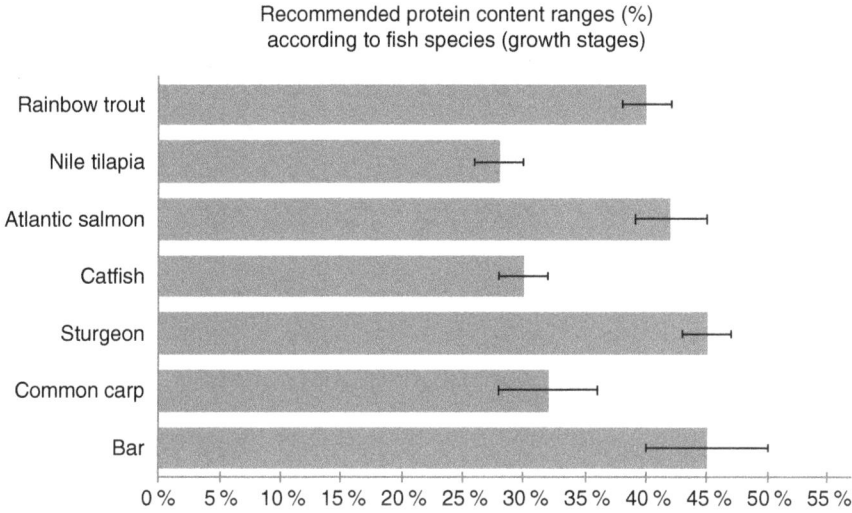

Figure 2-2. Ranges of protein levels (in %) recommended in the diets of different fish species (Pierre Foucard, ITAVI; based on data from the National Research Council, 2011).

population of fish over a given period. TCS is linked to the genetic potential of the species and its stage of domestication, as well as to rearing conditions (water quality, rearing density, presence or absence of sources of stress).

It is calculated as follows: $TCS(\%masse\ corporelle\ /\ jour) = \frac{Ln(Mf) - Ln(Mi)}{t} \times 100$ where Mi and Mf are the average initial and final individual masses and t is the length of the period in days. Figure 2-3 shows a growth curve developed for rainbow trout, the predominant species in French fish farming. It highlights the importance of temperature on fish growth, a factor that has a strong impact on zootechnical performance depending on the thermal preference of the species in question.

10.2.2.4. RATIONING RATE (TR). The daily ration to be distributed is determined using the ration rate TR expressed as a percentage of the biomass, which can be calculated as follows: $TR\ (\%\ biomass) = TCS \times IC$ where TR = Ration rate, TCS = Specific growth rate, IC = Conversion index.

The TR is generally between 5 and 10% for juvenile fish, and between 1 and 3% for sub-adult and adult fish. It varies greatly depending on the species, development stage, feed composition and water temperature. It tends to decrease

as the fish grow. Evolutionary rationing tables (for the life cycle of fish at the fry or commercial stage and for different water temperatures) are available on a turnkey basis from commercial feed suppliers.

10.2.3. Fish farming discharges and their impact on water quality

Many water quality parameters need to be taken into account for the health of farmed fish: ammonia nitrogen, dissolved nitrites and nitrates, dissolved oxygen, suspended solids, temperature and pH. This part concerns water chemistry and will be detailed in Chapter 3. Figure 2-4 gives an idea of the main water quality parameters affecting fish health.

Metabolic discharges from fish are the main source of degradation of this water quality and it is useful to know how they appear and accumulate in order to understand the issues involved in eliminating them.

10.2.3.1. FISH METABOLISM. Fish derive their energy from the consumption of complex molecules. These molecules are metabolised by the animal to ensure growth and various other physiological functions. The efficiency of these biological reactions is less than 100%, which

(A)

(B)

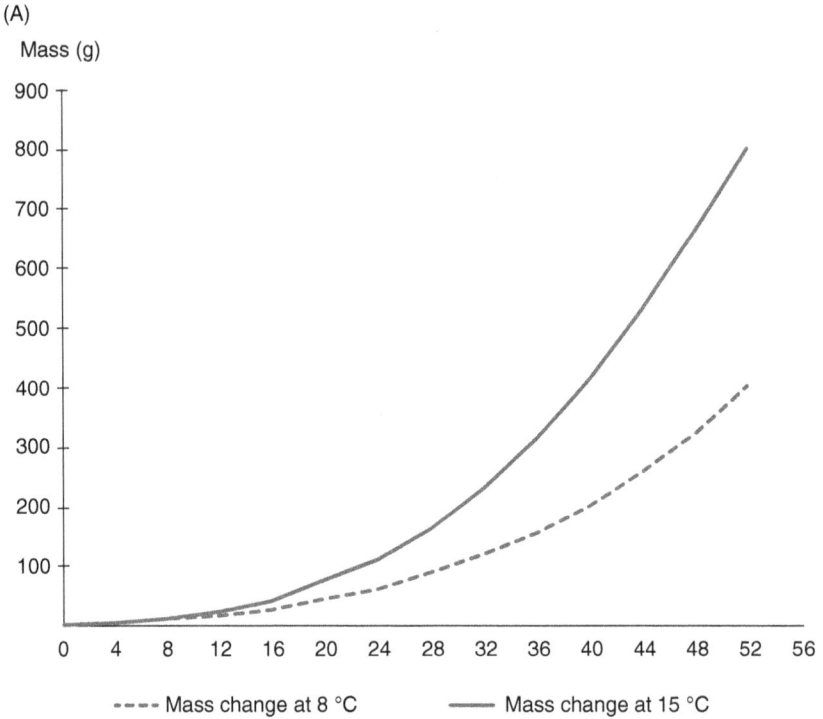

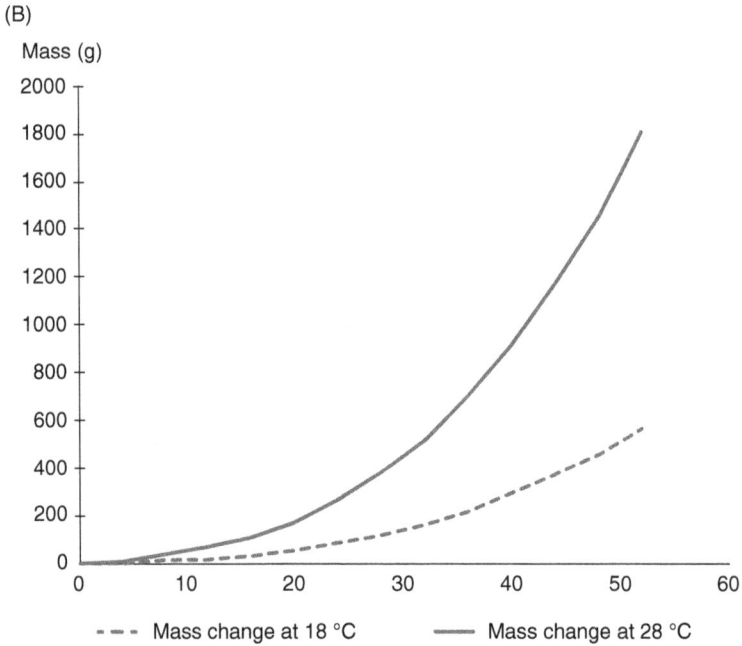

Figure 2-3. A. Growth of rainbow trout *Oncorhynchus mykiss* in freshwater at 8°C and 15°C. B. Growth of tilapia *Oreochromis aureus* in freshwater at 18°C and 28°C. (after Kaushik, 1999)

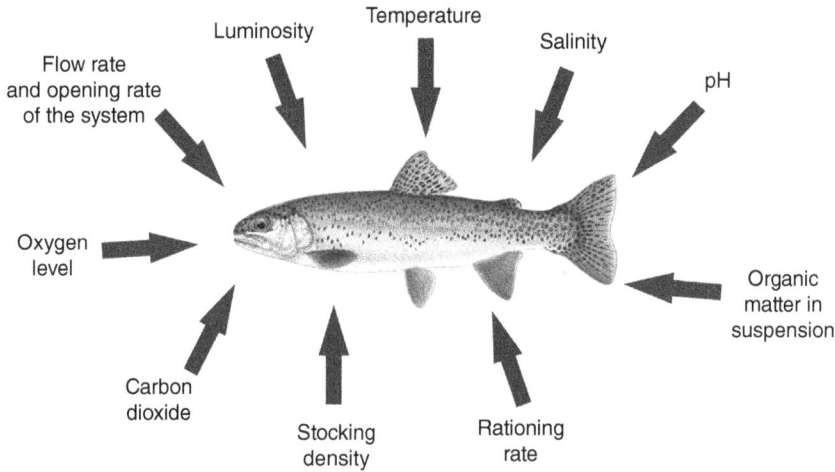

Figure 2-4. Main water quality parameters affecting fish health (Somerville *et al.*, 2014).

means that by-products of metabolism are discharged in dissolved and particulate form. Furthermore, even if feed distribution is finely controlled, it is always possible that some of the feed is not consumed and is evacuated in the form of solid waste (Blancheton *et al.*, 2004). The method, frequency and time of feed distribution, the duration of each meal and the quantity of feed delivered are all points that need to be controlled to avoid wastage. The quantity of faeces produced depends on the digestibility of the feed, which varies according to the nature of the nutrients, the quality of the feed and the stage of rearing. It is generally accepted that around 1/3 of the nitrogen released by fish is in solid form, with the remaining 2/3 in dissolved form. For phosphorus, 2/3 is in solid form and 1/3 in dissolved form (Kaushik, 1998a; Dosdat *et al.*, 1996 in Roque d'Orbcastel, 2008).

10.2.3.2. SOLID COMPOUNDS. Excrement and un-eaten feed make up the solid fraction of fish waste. The chemical composition and physical characteristics (size, density, hydration, shock resistance and fragmentation) of faeces vary according to the type of feed, fish species and development stage (Blancheton *et al.*, 2004). The size and composition of the faeces also depend on the residence time in the system, the presence or absence of bubbling in the fish ponds, the mechanical filtration method and the water renewal rate in the fish farming system, classically expressed as "litres of water per kilo of feed".

10.2.3.3. DISSOLVED COMPOUNDS. They are the result of excretion, mainly via the gills (80%) and the kidney (20%). The main dissolved metabolites are CO_2 for carbon, ammoniacal nitrogen (NH_3 and NH_4^+) and urea ($CO[NH_2]_2$) for nitrogen, and orthophosphates (PO_4^{3-}) for phosphorus. These molecules can contribute to the eutrophication of waters receiving aquaculture effluents (Martins, 2010; Edwards, 2015). Ammonia excretion pathways in freshwater fish differ from those in marine fish, due to different osmoregulatory mechanisms. Freshwater fish (hypertonic) urinate to maintain a low blood concentration of solutes, whereas marine fish (hypotonic) balance the high blood concentration by absorbing seawater and limiting urine excretion. As a result, freshwater fish provide a higher NH_4 generation rate[+] than marine fish at the same ration rate.

10.2.3.4. THE NEED FOR WATER FILTRATION. Effluent purification is a major issue in recirculating aquaculture systems. They generally combine a system for collecting and isolating the solids from fish farm waste and a biological treatment system for certain compounds dissolved in the water, using specialised bacteria (Roque d'Orbcastel, 2008).

If solid waste is not disposed of efficiently, it will stimulate the appearance of heterotrophic bacteria, which consume soluble organic compounds and oxygen, and are likely to compete with nitrifying bacteria, causing a rapid drop in the level of dissolved oxygen in the water (Rakocy *et al.*, 2006), and increasing the level of ammonia and hydrogen sulphide in the water. This can rapidly affect the health of fish (Pambrun, 2005) and lead to mortality. In addition, the accumulation of organic matter encourages the development of undesirable algae and bacteria, contributing to the overall clogging of the system and potentially reducing the flow rate (Timmons and Ebeling, 2007; Lennard, 2018). The quality of this mechanical filtration stage depends on the quality of biological filtration.

There are preconceived ideas about the ability of the system's bacteria to break down solid particles in aquaponics: while this may be true in a small-scale system, with a plant culture medium such as "clay balls" possibly colonised by earthworms, the reality is quite different on a large scale. It is vital to focus on a filtration system that is efficient, robust and durable over time. Sikawa and Yakupitiyage (2010) found that filtration of solid particles had a positive impact on plant growth compared with a control with unfiltered water. This can be explained by the fact that excessive organic matter leads to asphyxiation of the environment and the release of ammoniacal nitrogen, two undesirable phenomena in aquaculture and hydroponics.

10.3. Technologies used in recirculating aquaculture (RAS)

10.3.1. Mechanical filtration for coarse particles

The principle is to create a physical barrier, using granular materials (sand for example) or fine mesh (screening), through which the water to be treated flows to remove suspended solids (which we will call "SS"). It is important to know the flow rate of the circulating water and its TSS concentration in order to plan the most appropriate cleaning process; this can be calculated using mathematical models.

Different mechanical filtration technologies can be used alone or in combination, depending on the size of the particles to be captured and the purification efficiency objective: gravity or centrifugal decantation, sand and pressure filtration, filtration using microtamis or granular or porous materials (Acierno *et al.*, 2006; Timmons

and Ebeling, 2007). The choice between passive sedimentation and mechanical filtration depends on the degree of intensification of the farm. The lower the degree of intensification of the system and the lower the volume of water to be treated, the more appropriate it is to use sedimentation because of the reduced maintenance and costs, provided that sufficient water retention time is maintained in the decanter. The intensification of production (high stocking densities and higher feeding rates), the higher volume of water and the choice of more sensitive species make the use of drum filters more suitable and safer.

Rotary or drum filters are highly efficient systems for trapping solid matter. Fully automated, they are now the most widely used industrial recirculation systems. A filter of this type can retain between 50 and 90% of the particulate matter produced by fish (Roque d'Orbcastel, 2008). The principle is simple: the liquid to be filtered is poured into a rotating drum with meshes around its periphery (Figure 2-5). Impurities with a diameter larger than the mesh size are trapped against the inside face of the filter plates. Filtration grids can have a porosity of between 10 and 500 µm, but the main range is between 40 and 80 µm. The drum rotates slowly (3 to 8 rpm) and draws impurities out of the water, trapping them in the mesh. A variable-level sensor detects clogged filter cloth and sends a signal to the control panel to backwash the cloth. A rinsing ramp, located at the top of the drum, then sends pressurised water over the plates to remove impurities in an outlet channel.

Figure 2-5. Mechanical rotary filter. (Pierre Foucard, ITAVI)

This backwash water, loaded with organic matter, can then be more easily treated, by decantation for example, in order to evacuate compact sludge.

The choice of filtration system is an essential question to ask when sizing an aquaponics system. Table 2-1 gives a non-exhaustive list of the advantages and disadvantages of different technologies, the possible treatment flow rates and an estimate of the cost.

Mechanical filtration techniques based on gravity require a large surface area and involve maintenance; they are only effective in removing large settleable particles (> 100 μm), but are generally simple and inexpensive to build. Filtration techniques based on fine filtration often require more elaborate and automated equipment, which may be more expensive to purchase, but require much less floor space and are generally self-cleaning, although they do require some maintenance. Another possibility is to combine different filtration systems, which can be complementary: for example, a cylindrical-conical decanter combined with a drum filter will be more efficient than a drum filter alone. Even if they operate continuously, these systems are not 100% efficient, which means that over time fine particles (< 30 μm) will accumulate in the system. Some of these particles will eventually mineralise, and the rest will have to be eliminated by other means in order to avoid health problems for the fish, but also clogging of the biological filter and excessive consumption of oxygen as a result of their degradation by heterotrophic bacteria, the presence of which we are trying to limit.

Table 2-1. Advantages and disadvantages of different mechanical filtration technologies (Thorarinsdottir et al., 2015).

Type	Flow rate to be treated (m³/h)	Estimated cost	Benefits	Disadvantages
Simple decanter	5	1000	Low maintenance No electricity consumption	Takes up a lot of space Limited effectiveness for the finest particles
Cyclonic decanter	15 à 30	1200	Low maintenance No electricity consumption	Need for a large storage volume to ensure sufficient retention time
Ball filter	10	3000	Compact	Power consumption
	23	8000	Suitable for small to medium-sized farms	Maintenance required
	45	12,000		Frequent clogging
	68	20,000		Need to backwash regularly (water consumption), at highly variable frequency depending on the load of solid particles
Sand filter	10	700	Efficient and ideal for large farms	
	22	1200		
Rotary drum filter	30	5200	Efficient and ideal for large farms, low maintenance thanks to automated backwashing	Power consumption Need to clean filter grids periodically with hydrochloric acid or hydrogen peroxide
	90	7000		
	140	9000		

10.3.2. Elimination of fine organic particles

There are three main ways of countering the risk of fine particles accumulating in an ASR system:

– diluting the water in the system with new water. This is still the simplest and least expensive technique, but involves a slight consumption of water, from 100 to 1,000 litres of new water per kg of feed distributed, depending on the case;

– ozonation, which consists of chemically oxidising particles while agglomerating them in the same way as a flocculant, and which also makes it possible to destroy a very large number of bacteria, viruses, fungi and protozoan parasites of fish, and also to oxidise nitrite into nitrate. This technique requires a great deal of expertise and involves a significant increase in production costs (Timmons and Ebeling, 2007);

– foaming/skimming. *Skimmers* or *foam fractioners* create an air/water emulsion in a specific reactor, producing foam that rises to the surface where it is collected in a receptacle. The foam drains away molecules that are not very soluble in water (fats, proteins, fine particulate matter, etc.). This technique has long been used as a separative technique in various industrial fields, such as wastewater treatment, odour elimination and the harvesting of micro-algae and micro-organisms. It works well in salt water but is not very effective in fresh water.

10.3.3. Storage and recovery of sludge

On a large scale, both aquaponics and aquaculture are faced with the same problems when it comes to the production of livestock effluents. While it is clear that dissolved effluents can be used directly by plants, it is also relevant to consider what happens to solid effluents. In aquaculture, 30 to 70% of the phosphorus added to the system during fish feeding is lost in the form of solid excrement, a large proportion of which is filtered by settling tanks or drum filters (Schneider *et al.*, 2005). These phosphorus discharges have a eutrophying impact on the aquatic environment. Aquaponics is intended to

be a production system with a 'zero waste' objective, with maximum recycling of fish nutrients to produce plant biomass. Dalsgaard and Pedersen (2011) assessed solid and dissolved waste from a recirculating rainbow trout farm in Norway, and showed that 48% of ingested nitrogen (N) was recovered in water, with total ammonia nitrogen (TAN) making up 64-79%, and 7% in solids. In comparison, 1% of ingested phopshore (P) was recovered from water and 43% from solids. This highlights the importance of treating solid waste, which is a major source of nutrients for aquaponics plants.

Backwash water from mechanical drum filters has a variable dryness level of between 0.1% and 0.2% (Bergheim, 1998). It is generally left to settle in a storage tank, where it reaches a dryness level of 4% at best (Marcotte, 2007). A dryness of 13% to 15% is necessary to ensure that the sludge holds its shape and is easy to transport. Belt filter systems or geotextile bags (Figure 2-6) can be used to dewater the sludge to a suitable level, thereby reducing its volume and making it less expensive to transport (Aquinove, 1995). The water resulting from the leaching of

Figure 2-6. Filtration of sludge in a geotextile bag (Pierre Foucard, ITAVI)

this waste is rich in inorganic nitrogen and dissolved orthophosphates and can also be used again for the plant compartment in the case of aquaponic structures (Rakocy *et al.*, 2006).

An alternative way of using effluent is composting, which allows the sludge to be exported to market-gardening land that requires this product for soil improvement and structuring. Compost made from fish farm sludge is a high-performance fertiliser (Danaher *et al.*, 2013; Pantanella *et al.*, 2011a), but often needs to be balanced in terms of nitrogen, as the NPK ratio is often unbalanced. Carbon addition is also desirable (straw, sawdust, plant shavings, fine dry wood branches). There are also lombrifiltration processes (Figure 2-7), originally used to treat municipal water. This is provided by an "active layer" made up of wood chips, arranged in layers of different grain sizes. On the first layer, an inoculum of earthworms (species *Eisenia foetida*, *Eisenia andrei*, *Lumbricus rubellus*, etc.) is added with its rearing medium. This inoculum is introduced after the initial watering of the earthworm filter. The top layer is sprayed alternately by a mobile boom fitted with nozzles, and the sieved waste water is sprayed as rain. Thanks to this fine sprinkling, the water is spread over the entire surface, then percolates throughout the active layer. Finally, thanks to a bed of siliceous gravel, the lombrifiltrated water is drained. If properly designed, these systems can reduce organic matter by up to 95%, and recover

nutrient-rich leachate that can be used to enrich the plant nutrient solution for aquaponics (Bajsa, 2003, in Goddek *et al.*, 2015).

Another recovery method under study is the mineralisation/biodigestion of fish farm sludge by aerobic heterotrophic bacteria (e.g. *Lactobacillus plantarum*) in 'mineralisation tanks' (Jung *et al.*, 2011 in Goddek *et al.*, 2015). Mineralisation allows solid particles to be treated, broken down and mineralised in order to solubilise the nutrients trapped in the sludge, in a loop external to the fish farming system where the effluent to be treated is lightly aerated and stirred (1 to 2 mg/l of oxygen) to allow aerobic heterotrophic bacteria to develop. In its simplest form, a mineralisation tank is a reservoir containing charged water and a bubbler for aeration (Figure 2-8).

On a daily to weekly basis, the solid waste from the mechanical settling treatments is added to the mineralisation tank, where aeration is constant. At the same time, clarified water

Figure 2-7. Lombrifiltration system installed at the Lycée de la Canourgue in 2016 and tested as part of the APIVA® project for the degradation of fish farm sludge. (Catherine Lejolivet, EPLEFPA de Lozère)

Figure 2-8. Fish sludge mineralisation system installed at the ASTREDHOR-RATHO horticultural station (Rhône-Alpes) in 2018 and tested as part of the APIVA® project for the degradation of fish sludge. (Pierre Foucard, ITAVI)

containing dissolved nutrients is removed from the mineralisation tank (after the bubblers have been switched off for an appropriate time to allow the particles to settle) and can be reintroduced directly or *afterwards* into the aquaponics system (Lennard, 2012). The sludge will therefore not be completely 'liquefied' and 'mineralised', as an insoluble fraction will always remain: the primary objective for application in aquaponics is to recover the supernatant from the aerobic mineralisation treatment in order to reuse it as a source of minerals for plant production, while the stabilised residual sludge can subsequently be filtered and dehydrated for another application (composting/spreading), with the advantage of being chemically stabilised (Lennard, 2012). It is also possible to mineralise fish farm sludge under anaerobic conditions, but aerobic mineralisation works more quickly (Wallace and Knight, 2006 in Lennard, 2012). In addition, aerobic mineralisation presents fewer risks in use (no methane and ammonia production) and fewer operational problems. The potential for recycling nutrients from fish farm sludge was assessed in the study by Monsees *et al.* (2017b), with a comparative approach between aerobic and anaerobic sludge mineralisation. The nutrient 'remobilisation' process was thus investigated, with monitoring of nutrients $N-NO_3^-$, $N-NO_2^-$, TAN, PO_4^{3-}, K^+, Mg^{2+} and Fe^{2+} in sludge as well as the C:N ratio, total solids (TS) and suspended solids (TSS), all over 14 days of incubation. Under aerobic conditions, the concentration of dissolved phosphorus in the treated fish sludge was increased by 330% (from 9.4 ± 0.7 mg/l to 29.7 ± 2.1 mg/l), while it remained unchanged under anaerobic conditions. This demonstrates the benefits of aerobic mineralisation compared with anaerobic mineralisation in terms of the recovery of certain nutrients such as phosphorus. Both treatments also showed a potential for potassium remobilisation, with an increase of 31% for the aerobic treatment. It seems possible to improve these results with longer incubation times (Monsees *et al.*, 2017b). A sharp increase in ammonia levels was also demonstrated under anaerobic conditions (Monsees *et al.*, 2017b; Goddek *et al.*, 2016b), which shows that this technique is less suitable than aerobic biodigestion, as it is undesirable to incorporate nitrogen in ammoniacal form into an aquaponic system.

10.3.4. Biological filtration for the treatment of ammoniacal nitrogen

In recirculated aquaculture, the accumulation of dissolved ammoniacal nitrogen and its derivatives in effluents can be toxic or even lethal for fish, and is therefore a major concern. This is why recirculating systems use biological filtration of the farm water under aerobic conditions, based on the bacterial nitrification reaction, i.e. the oxidation of ammoniacal nitrogen produced by the fish into nitrates, a compound that is not very toxic to fish. The biological filter is positioned just upstream of the rearing tanks and downstream of the mechanical filter, which is itself positioned after the rearing tanks, in order to prevent the deposition of particulate organic matter in the biological filter as much as possible.

10.3.4.1. PRINCIPLE OF A FLUIDISED BED BIOFILTER. There are various types of biological filter, of which fluidised bed biofilter systems are the most efficient: The principle is to create turbulence in a bed of media by injecting air, thereby optimising the efficiency of bacterial nitrification and allowing the degassing of certain gases such as CO_2 (carbon dioxide), N_2 (nitrogen), NH_3 (ammonia) and H_2S (hydrogen sulphide), while oxygenating the farm water. A fluidised bed biofilter generally consists of a cylindrical tank filled with plastic media (*curlers*) to allow bacterial colonisation (Figure 2-9). A wide variety of media can be used, characterised by their specific surface area expressed in surface/volume units (200 to

Figure 2-9. Media mixing in a mechanical filter (Pierre Foucard, ITAVI)

5,000 m²/m³) and the material of which they are made: ceramic, polyester, polyethylene, etc.

10.3.4.2. BACTERIAL PLAYERS AND THE PRINCIPLE OF NITRIFICATION. Several strict aerobic bacterial genera of the chemolithoautotrophic type carry out nitrification, a natural reaction that forms part of the nitrogen cycle. Bacteria of the *Nitrosomonas* and *Nitrobacter* types respectively carry out the two distinct stages of nitrification, namely nitritation - from ammonium ion (NH_4^+) to nitrite (NO_2^-) - and nitration - from nitrite (NO_2^-) to nitrate (NO_3^-), chemical transformations that provide the bacteria with a source of energy (Figure 2-10). At the same time, the bacteria need carbonate ions as a carbon source and oxygen as an electron acceptor, two growth factors necessary for carrying out various metabolic functions.

The bacterial species *Nitrosomonas* and *Nitrobacter* are the best known and studied for their interest in the nitrification reaction, but many others may be involved in the process: *Nitrosococcus, Nitrosospira, Nitrosolobus, Nitrosovibrio, Nitrococcus, Nitrospira, Nitrospina* (Wongkiew, 2017b). During the degradation of ammonia by nitrifying bacteria, 1 g of N-NH_4^+ is degraded with 4.57 g of oxygen (Losordo *et al.*, 1994; Pambrun, 2005), resulting in the consumption of 7.14 g of alkalinity (Timmons and Ebeling, 2007) and the production of 5.93 g of CO_2. These data are important for knowing how to size a biological filter.

10.3.4.3. INSTALLATION OF A BIOFILTER ON A FLUIDISED BED. Nitrifying bacteria must be able to colonise supports or substrates in order to develop optimally (gravel, sand, synthetic media, etc.). Under optimum conditions of temperature (20-25°C) and pH (7.5 to 9, ideally 8), *Nitrosomonas* can double its population every 7 hours *and Nitrobacter* every 13 hours. This difference in generation time induces a peak in NO_2(-) nitrites during biofilter start-up (Figure 2-11).

The biofilters are preferably activated at a temperature above 20°C and at a pH between 7.5 and 8.5, before the fish are introduced into the rearing system. Activation usually takes place in a static phase. One possible method is to add ammonium chloride (NH_4Cl) at a dose of 4 to 10 mg/l, along with a carbon source at a similar concentration (molasses, for example) to

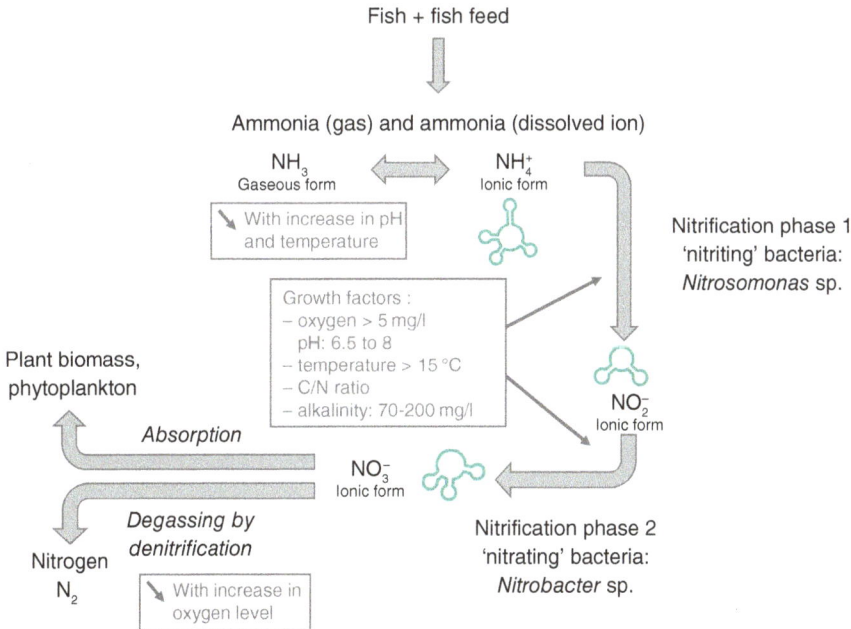

Figure 2-10. Nitrogen cycle (Pierre Foucard, ITAVI)

Concentration evolution
of nitrogenous form (mg/l)

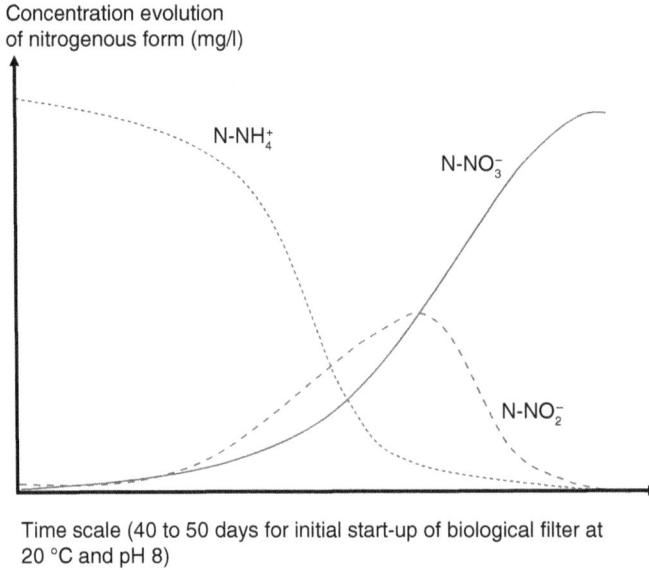

Time scale (40 to 50 days for initial start-up of biological filter at
20 °C and pH 8)

Figure 2-11. Illustration of the nitrite peak that occurs during seeding of a biological filter (Pierre Foucard, ITAVI)

stimulate the activity of nitrifying bacteria naturally present in the medium, or previously inoculated with commercial solutions (Roque d'Orbcastel, 2009). When using water from the drinking water network, it is advisable to use a commercially available inoculant of nitrifying bacteria; this should be done after leaving the water in the system under agitation for a few hours to facilitate the degassing of the chlorine.

Activation is complete when the NH_4^+ and NO_2^- levels drop to below 1 and 0.6 mg/l respectively, and the NO_3^- concentration rises. For a biofilter to start working, a variable timeframe of between 30 and 50 days is required; you need to wait at least 20 days after the seeding phase to start seeing clear changes in the physico-chemical parameters of the water. After 6 months, the biofilter has reached its maximum purification potential.

Ultraviolet radiation is harmful to bacteria during initial colony formation, so it is a good idea to cover the biofilter with an opaque tarpaulin during start-up, without creating an anaerobic environment. After two or three weeks, the bacteria have settled on the surface of the media and UV is no longer considered a threat (Somerville et al., 2014).

10.3.5. Aeration and/or oxygenation in recirculated systems

Oxygen is a major concern in aquaculture, just as it is in plant production. Fish extract oxygen from the water using their gills, an organ where gas exchanges are facilitated by a large exchange surface and a short distance between the water and the bloodstream. Blood leaving the heart passes directly through this gill system. The oxygenated blood is then distributed to all the organs. Gas exchange between the fish and the aquatic environment is maximised by counter-current circulation between the blood and the water. During the respiration process, fish release CO_2.

Oxygen content in water is expressed either as a concentration (mg/l or ppm) or as a percentage of saturation (% S). The saturation percentage gives an indication of the degree of equilibrium (for oxygen) between air and water. When % S < 100%, the water is oxygen undersaturated, and when % S > 100%, the water is oxygen supersaturated.

Oxygen requirements in aquaculture depend on many factors (Belaud, 1996):

–   the species. For example, the needs of trout or sea bass are greater than those of carp;

crustaceans and molluscs consume less oxygen than fish;
- the strain. Metabolically powerful strains can consume more oxygen, which increases their growth potential;
- temperature. The more the temperature rises, the more the requirements increase;
- mass/age. Juveniles consume more oxygen than adults for the same biomass;
- the availability of oxygen in the water. Consumption decreases when there is a shortage of oxygen in the surrounding environment;
- muscular activity. An active or excited fish consumes much more oxygen;
- digestion. After a meal, a fish consumes more oxygen;
- sex and state of sexual maturation;
- stress. The latter results from unsuitable rearing conditions, noise, vibrations, disease, etc.

Atmospheric aeration increases the concentration of oxygen in water up to a certain limit imposed by the partial pressure of air in the atmosphere (150 mmHg), but also by the solubility of oxygen in water. This solubility is limited by a threshold value that decreases with increasing atmospheric pressure (and therefore altitude), temperature and salinity. Up to a stocking density of 50 kg/m^3, an aeration system (immersed such as a blower, emerged such as a packed column, or even waterfalls when the terrain is suitable) may be sufficient to meet the fish's oxygen requirements, provided that the flow rate and water renewal rate in the system are adequate (Belaud, 1996). When stocking densities in excess of 50 kg/m^3 are used, an oxygenation system is required to allow the water to be slightly supersaturated with oxygen. As a result, intensive recirculating systems usually incorporate water oxygenation systems of varying complexity and efficiency (from 30% to 95%), such as U-tubes or the air lift used in 'Danish' type recirculating systems (Roque d'Orbcastel, 2008), bicones, jet platforms, or simply the injection of pressurised air directly into the hydraulic circuit (Belaud, 1996). The design of RAS oxygenation systems also takes into account the need to supply oxygen to the biofilter. Dissolved CO_2 released into the water by fish can become toxic in high concentrations (> 20 mg/l) (Kinkelin et al., 1985). The quantity of CO_2 in the water

can be adjusted by means of a degassing unit (cascades, aerator, aeration column, etc.). Aeration systems are often coupled with the degassing function.

10.4. The importance of the recirculation rate

The recirculation rate corresponds to the number of times the water in the rearing tanks is changed per hour. It is calculated by dividing the flow rate through the system (expressed in cubic metres of water per hour) by the storage volume in the rearing tanks. It is important to understand that we are talking about a movement of water and not a replacement of rearing water by 'new' water. The rate of recirculation will maximise oxygenation of the water and the purification efficiency of the mechanical filter. The more the water passes through the mesh of a mechanical filter, the less time organic particles will have to accumulate and degrade within the rearing water, and the less oxygen will be consumed by the bacterial activity linked to this degradation, leaving more oxygen for the fish. Similarly, the faster the water circulates, the more effective the oxygenation process by cascading and/or bubbling and/or oxygenation will be. A recirculation rate of at least two changes per hour is considered to be necessary in RAS.

10.5. The importance of the water renewal rate or "opening rate"

Even in a recirculating system, it is necessary to add new water every day to replace and dilute part of the water in the system, not only to compensate for water lost through evaporation and evapotranspiration, but also to avoid the accumulation of fine organic particles (< 60 μm), or of certain minerals that can be harmful to fish in excessively high doses, such as iron, copper, zinc, or nitrates, the toxicity of which in high concentrations is not yet fully understood and is highly species-dependent. We are also trying to avoid excessive accumulation of sodium (Na) - fish feed contains a certain amount of salt - which can rapidly become toxic for many plants above 30 mg/l, with very marked differences in tolerance depending on the species.

Water renewal determines what is known as the system's "opening rate", which can be expressed in different ways:

- the ratio of the volume of new water (m^3)/volume of water contained in the system (m^3), which generally varies between 1 and 10%;
- the ratio of volume of new water (m^3)/quantity of feed distributed (kg), which generally varies between 0.2 and 1 m^3/kg of feed.

10.6. Restriction of chemical inputs for fish farming

10.6.1. Zoosanitary treatments in fish farming

The use of antibiotic, antiparasitic or antifungal products must be limited in aquaponics. In the absence of precise knowledge on the subject, we can only assume that these compounds are absorbed and concentrated by the plants, which could prove problematic for products intended for human consumption. The nitrifying bacteria in the biofilter are just as likely to be affected by these substances (Zhang *et al.*, 2017). This implies increased monitoring of fish farming, the difficulty of managing high stocking densities, and the use of prophylactic measures such as probiotics/prebiotics as well as systematic disinfection of the renewal water. Decoupled aquaponics systems do, however, offer flexibility for the fish farming compartment, allowing treatment to be used if absolutely necessary, by decoupling the system from the plant part for a few days, while any traces of undesirable compounds are evacuated.

10.6.2. Diagnosis of the state of health of farmed fish

Among the advantages of recirculating fish farming technology, the issue of disease control is an important one. The impact of pathogens is considerably reduced in a closed circuit due to the use of a small quantity of new water, the quality of which can be controlled, in particular by installing a UV filter or using ozone. Fish pathogenic organisms generally come from the outside environment, in particular from renewal water. In an open circuit, water from a river is most often used, which naturally increases the

chances of a pathogenic organism of exogenous origin entering the tanks and reaching the fish. In a closed circuit, low water consumption means that you can make do with small quantities of new water, which is more easily treated with UV or ozone, reducing the risk of disease and therefore the use of curative medicinal products. Diseases can also occur as a result of fry entering in a poor state of health. It is therefore important to use the services of a professional who can provide veterinary certificates (certificate of disease-free status for the species being reared), or to integrate hatchery activities into the rearing process to ensure an in-house quality approach. In addition to exogenous pathogenic bacteria, physico-chemical parameters can have a significant impact on fish health (oxygen, pH, temperature, etc.), especially in the case of rapid fluctuations, and can encourage the appearance of pathologies.

The well-being and state of health of fish can easily be monitored through their behaviour: healthy fish swim without lethargy or overexcitement, have a good appetite and do not show marks - scratches, wounds, missing scales - along their body. Certain signs of discomfort are easy to recognise and can usually indicate a pathology, the presence of parasites or a state of stress of some kind: hyperventilation, jerky swimming on the surface or standing still for long periods, immobile position at the bottom of tanks, fins stuck together, presence of wounds, swollen eyes or belly, damaged fins, necroses, ulcers, excessive jumping by the fish, tendency of the fish to rub against the walls of the tanks. Fish that feed little or not at all can indicate poor stock health, poor water quality or overfeeding. All these signs, potentially indicative of a deteriorated state of health or the onset of disease, must be validated and confirmed under the responsibility of a veterinary practitioner specialising in aquaculture.

11. The hydroponic compartment: the plants

11.1. Possible plant species for aquaponics

Many plant species can be grown in an aquaponics system. Plants with low to medium nutrient requirements are the easiest to grow: lettuces and other salads, Chinese cabbage, chard, spinach, broccoli, and herbs such as chives, basil, watercress, mint and rocket (Diver, 2006). Other plant species that have a higher nutrient demand - and which will be more suitable for intensive fish farming and additional supply of limiting nutrients - have also been tested: tomatoes, strawberries, peppers, cucumbers, beans, cauliflower, peas, squash, etc. (Savidov *et al.*, 2005). Figure 2-12 - the result of a statistical study carried out worldwide on the diversity of aquaponic farming practices - gives an idea of the plant species most commonly used (Love *et al.*, 2014).

11.2. Basic crop science

11.2.1. Crop density

Working above ground means that there is no root competition (Jones, 2005). This means that crop density can be higher than in the field, and is limited only by the leaves' access to light and the diameter of the plants at harvest size. Typically, lettuces can be grown at 16-20 plants/m², while most herbs can be grown at 32-40 plants/m².

11.2.2. Plant nutrition: the importance of nutrients

Plants are autotrophic organisms, feeding themselves through a series of processes that enable them to absorb and assimilate the nutrients they need for their various physiological functions: growth, development and reproduction. Plants must absorb certain minerals through their roots if they are to survive.

In addition to carbon, oxygen and hydrogen, which are supplied by carbon dioxide and water during the photosynthesis process, 13 other elements are essential: macro-elements

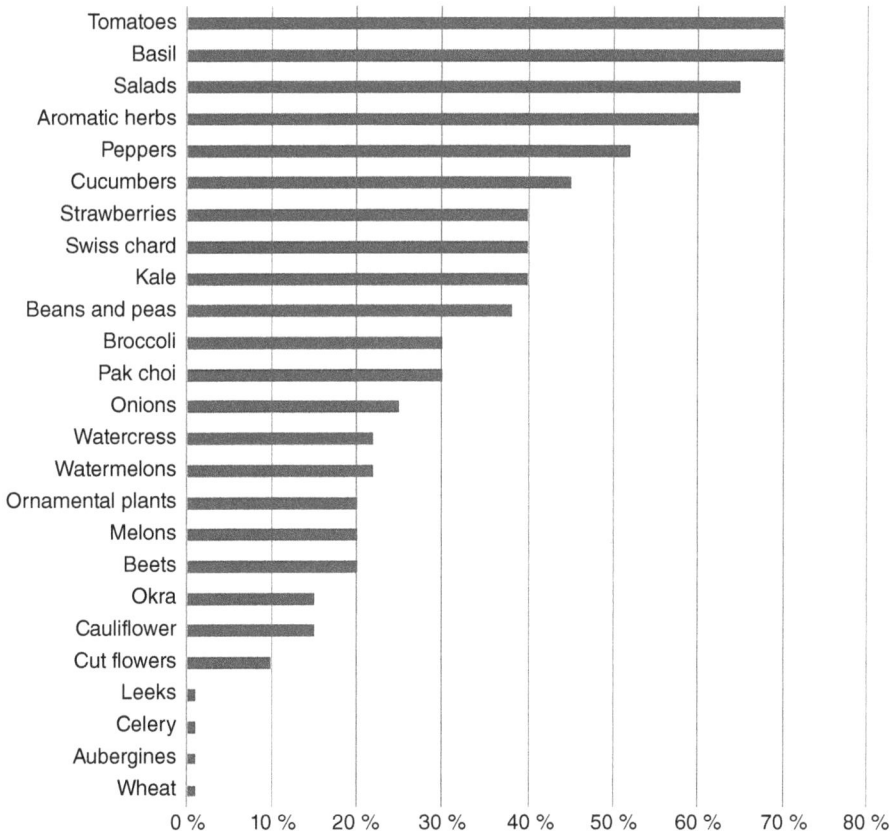

Figure 2-12. Plant species most commonly grown in aquaponics. (Love *et al.*, 2014)

such as nitrogen (N), phosphorus (P), potassium (K), calcium (Ca), magnesium (Mg), sulphur (S) and micro-elements such as chlorine (Cl), boron (B), iron (Fe), manganese (Mn), zinc (Zn), copper (Cu) and molybdenum (Mb).

Some elements are considered non-essential but sometimes 'beneficial' for certain plants in small quantities, such as chlorine (Cl), nickel (Ni), cobalt (Co), silicon (Si) and sodium (Na).

Plants are also said to be able to use free amino acids for growth (Somerville *et al.*, 2014), which is a good thing in aquaponics as decomposing fish feed (from losses during feeding phases) can potentially release proteins and amino acids into the environment.

Macro-elements are mainly - but not exclusively - involved in the structure of molecules. Requirements for micro-nutrients are relatively low, and they usually act as catalysts or regulators, for example of enzyme activation. Some macronutrients, such as calcium and magnesium, have a dual structural and regulatory role.

Nutrients need to be present in quantities that are in balance with the standards established for hydroponics, as excessive quantities of certain nutrients can lead to a drop in the bioavailability of other nutrients or even a toxicity threshold, as illustrated in figure 2-13.

Nutrient imbalances in plants can lead to biochemical and morphological disorders. Ca, Mg and Na in excessively high proportions interfere with the uptake of potassium and iron (Hochmuth, 2012). In addition, fruiting plants such as tomatoes need a lot of nitrates during the vegetative stage, as do most plants, while these same nitrates in too high a proportion may subsequently inhibit fruit development: a N/P/K ratio favourable to phosphorus and potassium is therefore necessary (Jones, 2005; Rakocy *et al.*, 2006). It is generally considered that too much calcium or magnesium can cause a potassium deficiency: K/Ca and K/Mg ratios should ideally be above 2:1 and below 10:1.

Table 2-2 shows the typical composition of nutrient solutions in hydroponics, with orders of magnitude of desirable concentrations for the various macro- and micronutrients. Each element is responsible for a specific physiological role, and any deficiency or excess can be detrimental to plant production (Jones, 2005; Osvalde, 2011).

In hydroponics, plants have access to nutrient solutions containing 100% mineral nutrients, dissolved in ionic form, whereas in aquaponics, the nutrient solution is made up of a mixture of organic and mineral molecules from the metabolism of the fish. There is very little in the literature about how plants consume organic molecules. Another major difference is the presence of a specific microflora in aquaponics,

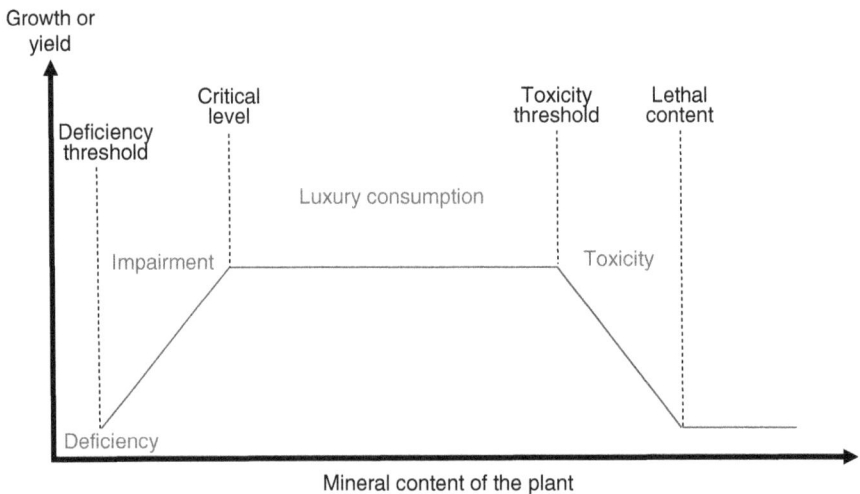

Figure 2-13. Relative effect of the mineral content of the nutrient solution on growth. (Pierre Foucard, adapted from CTIFL, 1995)

Table 2-2. Typical composition of nutrient solutions in hydroponics. (Pierre Foucard, ITAVI, adapted from Jones, 2014)

	Chemical element	Chemical symbol	Desired concentration range (mg/l)* (mg/l)	Dissolved form available to plants
Macronutrients	Nitrogen	N	100 to 200	$N\text{-}NO_3^-$, $N\text{-}NH_4^+$ (more than 95% in the $N\text{-}NO_3^-$ form)
	Phosphorus	P	30 to 80	$P\text{-}PO_4^{3-}$
	Potassium	K	100 to 200	K^+
	Calcium	Ca	200 to 300	Ca^{2+}
	Magnesium	Mg	30 to 80	Mg^{2+}
	Sulphur	S	70 to 150	SO_4^{2-}
Micronutrients	Boron	B	0.03	$B(OH)4^-$
	Copper	Cu	0.01 to 0.1	Cu^{2+}
	Iron	Fe	2 to 12	Fe^{2+}, Fe^{3+}
	Manganese	Mn	0.5 to 2	Mn^{2+}
	Molybdenum	Mo	0.05	MoO_4^{2-}
	Zinc	Zn	0.05 to 0.5	Zn^{2+}

*Expressed in relation to the element and not in relation to the dissolved form.

which does not exist in hydroponics where the media are often sterilised to avoid the development of bacteria or fungi that are pathogenic for plants (Lennard, 2018): the effects of this microflora potentially have beneficial effects on plant growth and the assimilation of organic molecules (Goddek et al., 2015), as well as on the solubilisation of particulate phosphorus (change from the 'solid' form to the ionic form PO_4^{3-}).

Some authors believe that aquaponics offers similar or even higher yields than hydroponics, despite lower concentrations of mineral nutrients (Savidov, 2004; Rakocy et al., 2006). Compared with the nutrient solutions applied in hydroponics, fish farm effluents produced in a recirculated circuit generally contain 3 to 10 times less nitrogen and phosphorus and 5 to 10 times less potassium and magnesium. Nevertheless, Graber et al (2009) report tomato yields almost identical to those obtained in hydroponics, the most limiting element for plant growth being potassium, which is most often supplemented with potassium bicarbonate, used basically to correct the pH while providing a source of alkalinity to feed the bacteria in the biological filter. Table 2-3 lists the macro- and micronutrient values measured in the nutrient solution intended for plant irrigation as described in

several studies on aquaponics, comparing them with the composition of hydroponic nutrient solutions in order to highlight the significant differences that can exist between an aquaponic and a hydroponic solution.

According to the literature, calcium, potassium and iron are the three nutrients considered to be the most limiting in aquaponics, and sometimes need to be supplemented periodically to avoid deficiencies. According to some authors, tomatoes grown in aquaponics may also lack phosphorus, manganese, sulphur, zinc or boron (Rakocy et al., 2006; Roosta, 2011).

Studies carried out in aquaponics have measured fairly low phosphorus concentration ranges in the plant nutrient solution, between 1 and 17 mg/l $P\text{-}PO_4^{3-}$ (Goddek et al., 2015), whereas the recommended concentrations in hydroponics are around 40 to 60 mg/l $P\text{-}PO_4^{3-}$ (Goddek et al., 2015). This is rarely a problem for leafy vegetables and herbs, but can be limiting for growing plants with high requirements such as tomatoes or other fruiting vegetables. Phosphorus production is a major concern for the future of agriculture, given the growing demand for this input in mineral form, its non-renewable nature - which poses threats to its availability in the medium or long term - and the eutrophying

Table 2-3. Macro- and micronutrient concentrations (mg/l) in aquaponic irrigation solutions, compared with a conventional hydroponic solution (Pierre Foucard, ITAVI)

	pH	EC	N-NO$_3^-$	P-PO$_4^-$	K	Ca	Mg	SO$_4$-S	Na	Fe	Mn	Cu	Zn	Mo	B	Source
		mS/cm	mg/l							µg/l						
Aquaponics	7.4	0.7-0.8	42	8	45[a]	12	7			2500[b]	800	50	440	10	190	Rakocy et al., 2004
	7.1	0.7-0.9	26	15	64[a]	24	6	6	14	2500[b]	60	30	340	10	90	Rakocy et al., 2004
	8		20	10	48[a]											Al Hafedh et al., 2008
	5.6-7.3		20	17												Endut et al., 2010
	7.7		35	8	27	34				200		40	370			Roosta and Hamidpour, 2011
			137	9	106[a]	180	44		17							Pantanella et al., 2012
			46.6-52.4	7.1-8.5		12.7-19.1	6.8-8.5	9.2-12.3	5-74.3	1580-4330[a]	320-600	50-120[c]	111-190	140-410	240 to 600[c]	Delaide et al., 2016
	6.8-7.3	0.8-1.2	93.0	9.6	89[a]	123.0	9.2	28.0	27.0	20.0	10.0	10.0	70.0	1.0	47.0	ITAVI, 2016
	7.5-8.2	0.6-0.8	35.0	5.5	5.9	128.0	8.2	21.0	24.0	10.0	10.0	10.0	70.0	1.0	42.0	ITAVI, 2017
	6.8-7.5	0.7-0.9	65	17	55[a]	103	15.3	23	39.8	2000.0[b]	140.0[c]	272	455	109	153	ITAVI, 2018
			10.6	6.6	50.8	129.6	20.9			80		80	170	3	80	Bittsansky et al., 2016
	5.1-6.9		84	3.5	48	90	15			100						Nozzi et al., 2018
	5-7.3		62	1.9	35	74	11			1800[b]						
	5-6.5		82	28[d]	146[a]	74	32			2100[b]						
Hydroponics	5-6.2	1.5-3	100-200	30-60	100-200	150-300	35-60	50-330	< 20	3000-10000	500-3000	20-400	500-1000	10-200	100-500	Osvalde, 2011; Trejo-Téllez et al., 2012; Jones, 2014

[a]: potassium input
[b]: chelated iron input
[c]: microelement input
[d]: phosphorus input

effect it has on aquatic environments (Ragnars-dottir, 2011). It is therefore strategic to try to avoid the use of phosphorus from mining sources in aquaponics, in order to maintain a virtuous image that limits its impact on the environment. The process of aerobic biodigestion of sludge is being studied as a way of recovering the phosphorus contained in large quantities in organic form in fish farming sludge.

The composition of the new water plays an important role in whether or not deficiencies appear; 'hard' water will contain a lot of calcium and a deficiency will be highly unlikely, whereas acid water or even rainwater will have a very low calcium composition. Deficiencies in Ca, Fe, Mn, Zn, Cu, S and Bo first appear in young tissues, due to the non-mobility of these elements in the plant, whereas deficiencies in N, P, K and Mg appear in older tissues due to the mobility of these elements in the plant and their relocation in the youngest leaves when nutritional deficiencies occur.

11.2.3. Plant nutrition: the importance of light

Light has a major impact on photosynthesis by plants (Pramanik *et al.*, 2000). Sunlight is obviously the most efficient and least expensive source, but supplemental lighting can improve the situation in winter in temperate countries. Various technologies exist for lighting hydroponics: incandescent lamps, sodium lamps (HPS), halogens, fluorescent tubes, compact fluorescent lamps, LEDs, etc., with equally varied investment and energy costs. The intensity and duration of lighting are specific to each plant species, and need to be adapted according to the stage of growth. It is crucial to know and distinguish between two 'measures' of light: illuminance and photosynthetically active radiation:

– *Illuminance* can be defined as a quantity of lumens/m^2, which corresponds to lux, the SI (Système International) unit of illuminance. It corresponds to the density of photons falling on a given surface. It can be measured with a luxmeter. This definition implies that the closer the lamps are to the surfaces to be lit, the higher the luminous intensity (but the smaller the surface area lit). On sunny days, away from direct sunlight, illuminance is generally 20,000-50,000 lux.

Under overcast skies, incident light rarely exceeds 1,000 lux, and at dusk, around 10 lux. The light output to aim for for plants depends on the species, the stage of growth and the surface area to be illuminated. The lux is a unit of measurement that should be taken with great caution, as it quantifies the quantity of light perceived by a human being (wavelengths close to green/yellow). As a result, this system of measurement refers to human vision and not to the way in which plants can harness the energy of light. Although domestic lamps produce an illumination level suited to human comfort, they produce very mediocre growth results. The lux is therefore not very appropriate for estimating the effectiveness of lighting for photosynthesis. PAR, on the other hand, is a much more useful unit of measurement for greenhouse growers;

– *Photosynthetically* Active Radiation (PAR) corresponds to the wavelength range of light best suited to photosynthesis by plants. PAR, also known as PPFD (*Photosynthetic Photon Flux Density*) is expressed in $\mu mol/m^{(2)}/s$, and corresponds to the light energy received by the illuminated surface, i.e. the leaves of plants. Photosynthesis is a process that requires light energy and is best carried out in wavelengths between 400 and 700 nanometres, the range known as "visible light", which encompasses the electromagnetic spectrum from blue/violet to red. When visible light reaches the Earth, the different "objects" it hits absorb or reflect different wavelengths, producing a visible colour. The wavelength reflected by a surface will correspond to the colour that the object presents to our eyes. For example, most plants appear green because the chlorophyll in their cells reflects green light and absorbs virtually none. Sensitivity to wavelengths is not the same for every type of plant, and depends on the pigments involved in photosynthesis. The best-known and most active pigments are chlorophyll a and b, which are particularly sensitive to red and blue light. Blue light provides higher energy and a shorter wavelength than green or red light. Red light provides the smallest amount of energy in the visible spectrum. The light spectrum reaching plants must be balanced, with proportions of blue/violet and red/orange waves: red rays are more active for photosynthesis but induce excessive stem elongation, while blue light favours short internodes. Green plants do not therefore use the entire visible light spectrum, but only a well-defined range of wavelengths in which the pigments are active to light radiation. It is this range of waves of different colours used by plants that is known as "PAR light".

Plants have specific requirements in terms of the spectral composition of light. This will have an impact on the choice of lamps used for plant lighting if an artificial lighting strategy is adopted. These will need to convert as much electrical energy as possible into light that can be assimilated by the plant (PAR). Optimising the light spectrum, the PPFD and the photoperiod for each plant species and each stage of development are all research questions that are still in their infancy. The choice of a lighting system will therefore depend on a number of criteria:

–   luminous efficiency, i.e. the electrical energy converted into light energy;
–   PAR performance;
–   running costs (depending on the power used);
–   the life of the bulb.

According to Singh et al. (2014), recent developments in light source technologies have opened new perspectives for sustainable and highly efficient light input sources: LEDs. Since LED technology offers great flexibility in the way output light spectra can be designed, optimal adaptation of lighting conditions to the specific needs of plants can be achieved. LEDs offer a new energy-efficient approach to lighting buildings and greenhouses that can reduce the cost of production of vegetable and ornamental crops. Recent studies have shown that tomato growers using LED-type lighting can achieve yields equivalent to the results obtained with conventional horticultural lighting, and at a production cost 25% to 30% lower due to LEDs' high energy efficiency, low maintenance costs and longevity (Singh et al., 2014). These results need to be tempered in the light of another equally recent study (Nelson and Bugbee, 2014), which states that the most efficient HPS lamps and LEDs on the market have the same efficiency, but that LEDs have a cost in terms of "initial capital per photon delivered" that is five to ten times higher than HPS lamps. The cost of LEDs over 5 years would therefore be double that of HPS (Nelson and Bugbee, 2014). Despite contradictory results in the literature, LED lamps are the subject of very active research. Technical improvements and the falling cost of this lighting solution are making it increasingly attractive. The potential of LEDs is far from being fully explored, and more research is needed to study all the effects of this type of lighting on the physiology of different plants at different stages of growth or development.

11.2.4. Water quality

As with fish, this is a major factor in plant health. Different water quality parameters need to be taken into account in hydroponics: dissolved oxygen levels, temperature, pH and conductivity.

11.2.5. Growing media: role, characteristics and possibilities of substrates

In hydroponics and aquaponics, plant susbtrates form the support for the plant. Even in soilless cultivation, plants need a substrate to spread their first roots and maintain their equilibrium. These substrates are of no interest for plant nutrition because they are generally inert and very poor in nutrients. They will therefore not nourish the plant - that is the prerogative of the nutrient solution in soilless cultivation - but only help it to maintain an upright growth habit. According to Morel et al (2000), an ideal growing medium should be made up of a material with:

–   total porosity greater than 85%;
–   a dry density of 0.1 to 0.3;
–   an air retention capacity at pF1 (% of volume) greater than 20% to allow good aeration of the roots;
–   a water retention capacity at pF1 (% of volume) of 55 to 70%;
–   little or no cation exchange capacity;
–   the ability to buffer variations in salinity and pH when nutrient solution is added in order to limit electrical conductivity to a range of 600 to 2,000 μS/cm;

– negligible chemical reactivity so as not to affect the pH and composition of the nutrient solution.

A culture substrate must also be physically stable, non-deformable, easy to recycle, free from pathogenic germs and toxic substances, and have an acceptable cost (Morel *et al.*, 2000). Table 2-4 shows a selection of the main existing substrates.

11.3. Soilless culture methods that can be combined with the aquaculture compartment

According to Morard (1995), soilless cultivation is defined as "cultivation of plants carrying out their entire production cycle without their root system having been in contact with their natural environment, the soil". Soilless systems generally offer only limited space for root development. They are not suitable for growing large plants with extensive root systems. Three main hydroponic techniques are used in conjunction with aquaculture in existing aquaponic systems around the world: Deep Water Cultivation (DWC), Nutrient Film Technique (NFT) and Media Filled Growbeds (MFG). It should be noted that anything is conceivable, and that aquaponics is not confined to these three techniques, even though they are the ones most often cited in the literature. Other techniques may be of interest and deserve to be explored, such as sub-irrigation, percolation, vertical cultivation and aeroponics.

11.3.1. The rafting technique (DWC: Deep Water Culture)

This technique is the most widely used in large-scale aquaponics, as it makes it easier to rotate crops and plan production. The plants are grown on *rafts*, which are floating plates (usually made of extruded polystyrene, 40 to 50 mm thick) laid directly on the water (15 to 30 cm deep) in rectangular growing basins of varying dimensions, laid on the ground or raised to human height. Figure 2-14 shows a diagram of this growing technique, while Figure 2-15 illustrates lettuce grown in *rafts*.

The plants are supported by an inert substrate placed in culture pots, which in turn are placed through culture plates called "*rafts*", in holes provided for this purpose. The culture plates are pre-drilled with culture densities adapted to the targeted production strategy. This technique uses a continuous flow of water, with the plant roots constantly irrigated with well-oxygenated water. Once developed, the plant roots literally 'soak' in the water. In *raft* culture systems, oxygenation of the nutrient solution is managed by the flow rate circulating under the shelves; we generally aim to achieve a flow rate of between 0.25 and 1 water renewal per hour to ensure sufficient water movement to oxygenate the water. It is also possible to install air diffusers at the bottom of the *rafts* along their entire length.

The surface under the *rafts* and the space between the roots of the plants are potential habitats for nitrifying bacteria, but they are by no means sufficient, and a precisely dimensioned biological filtration compartment is needed upstream: it is not *a priori* conceivable to use these culture surfaces to treat the ammoniacal nitrogen produced by farmed fish. This technique has been tried and tested, but it sometimes poses problems for long-cycle crops, due to the organic sediments (non-filterable fine particles < 25 µm, plant waste) that settle to the bottom. If this solid waste is not properly managed, it could lead to an accumulation of sludge on the plant roots, or even asphyxiation of the environment. This is why it is preferable to opt for short crops and/or to ensure optimum mechanical filtration upstream, and to clean the bottom of the growing basins two or three times a year.

11.3.2. The NFT (Nutrient Film Technique) gutter cultivation technique

More commonly used in hydroponics, this technique is also used in aquaponics. Plants are grown on gutters (made of polyethylene, PVC, etc.), which are generally placed on supports that can be easily dismantled at ground level. A very thin film of water flows down each channel of the gutters, which have a slope of around 1%. A flow rate of 1 to 2 l/min is recommended (Jones, 2005). Figure 2-16 shows a diagram of

Table 2-4. Description and characteristics of different hydroponic substrates. (Morgan, 2003)

Substance	Features
Inorganic (inert)	
Rock wool	Non-toxic, sterile, very light when dry, reusable, excellent water retention capacity (80%), good aeration (17% air retention), no cation exchange or buffering capacity. Provides an ideal environment for roots to germinate seeds and for plants to grow over the long term.
Vermiculite	Porous, spongy, sterile, light, high water absorption capacity (5 times its own weight). Easily reaches its water saturation point. High cation exchange capacity.
Perlite	Siliceous, sterile, spongy, very light, free-draining, with no cation exchange or buffering capacity. When mixed with vermiculite, excellent germination medium. Its dust can cause respiratory irritation.
Fine gravel	Particle size varies from 5 to 15 mm in diameter. Free drainage. Low water retention capacity. High bulk density, an advantage that is sometimes perceived as a disadvantage. May require thorough washing and sterilisation before use.
Sand	Small rock grains of various sizes (ideally, from 0.06 to 2.5 mm in diameter). Varies in mineral composition. Can be contaminated with clay and silt particles, which must be removed before use in a hydroponic system. Low water retention capacity. High water density. Frequently added to organic media to make them heavier and improve drainage.
Expanded clay	Sterile, inert, pebbles from 1 to 18 mm, free drainage. Its physical structure allows water and nutrients to accumulate. Reusable if sterilised. Commonly used in hydroponic pots.
Pumice stone	Siliceous material of volcanic origin. Higher water retention capacity than sand. High porosity.
Slag	Porous volcanic rock. Fine quality slag is used in germination mixes. Very light. Higher water retention capacity than sand.
Polyurethane tiles	New material, 75% to 80% air space and 15% water retention capacity.
Organic (non-inert)	
Coconut fibre	The fine fibre is used for germination. Other forms: peat/coconut fibre, palm peat. Useful in capillary systems. High water and nutrient retention capacity. Can be mixed with perlite to form media with different water-holding capacities. Particle size can vary widely. Sodium contamination possible.
Peat	Used in seed germination mixes and potting soils. Reaches water saturation point quickly. Usually mixed with other materials to vary its physical and chemical properties.
Composted bark	Used in potting soils as a substitute for peat; particle size can vary; must be composted to reduce the toxic materials present in the original pine bark (*Pinus radiata*). Rich in manganese (Mn), can affect the azotic status of plants when first used. Prevents the development of root diseases.
Sawdust	Fresh uncomposted sawdust, with a medium to coarse texture, is ideal for short-term use. Adequate water-holding capacity and reasonable aeration. Decomposes quickly, which is a problem for long-term use. In addition, its source can seriously affect its acceptability.
Rice husk	Less well known and therefore little used. Its properties are similar to those of perlite. Excellent drainage. Average or low water retention capacity. Depending on its source, may contain chemical residues, requiring sterilisation before use.
Sphagnum moss	Common ingredient in many types of soilless media. Varies considerably in physical and chemical properties depending on its origin. Ideal for seed germination; excellent medium in screened pots used with NFT systems. High water retention capacity. Reaches saturation point easily. Provides gardeners with a degree of control over root diseases.
Worm castings/ Composts	Worm castings and composts, which vary considerably in their chemical composition and nutrient content, are used in organic hydroponic systems. They can become saturated with water. To alter their physical properties, they are mixed with other organic materials or with pumice, coarse sand or slag.

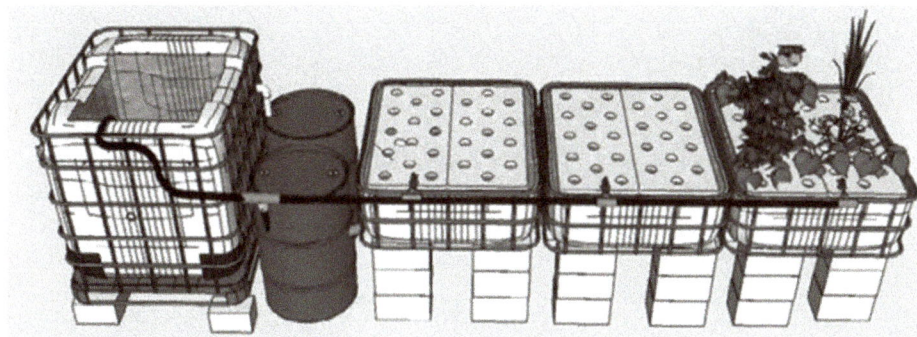

Farming tank
+ biological filtration

Raft cultivation system
(Deep Water Culture, DWC)

Figure 2-14. Schematic diagram of the principle of *raft* culture. (Somerville *et al.*, 2014)

Figure 2-15. Growing lettuce on extruded polystyrene *raft* plates (Pierre Foucard, ITAVI)

this growing technique, while Figure 2-17 illustrates lettuce grown on NFT gutters.

The plants are supported by an inert substrate placed in growing pots, which are themselves placed through the NFT growing troughs, in the holes provided for this purpose. The gutters are perforated at regular intervals along their length, and the distance between them can be adjusted. This technique uses a continuous flow of water, so the plant roots are constantly irrigated with well-oxygenated water. As with the DWC technique, the plant roots literally 'soak' in the water. In NFT cultivation systems, oxygenation of the nutrient solution is largely achieved by its movement through the gutters and by the large surface area exchanged between the water and the air. As with the *raft*

technique, this plant culture compartment is absolutely no substitute for a biological filtration compartment for fish effluent.

11.3.3. Growing on a bed of inert substrate (MFG: Media Filled Growbed)

This technique is most often used for small-scale, leisure aquaponics, where maximising production is not an objective. It can be used to grow a wide range of plants (including tomatoes, cucumbers and strawberries) and simply requires a growing tank or container of some kind filled with a neutral, inert substrate such as gravel/clay pebbles, etc., for the plant compartment, which acts as both a growing medium and a support for the plants. These media are continuously or discontinuously irrigated with a nutrient solution that provides the mineral salts essential for plant growth directly to the roots. This system can be used in two different ways:

– with a continuous flow of water as in a *raft* or NFT and simple evacuation of the water by an overflow, all with a sufficient flow to ensure good oxygenation of the water;

– by successive flooding and draining of the growing medium, also known as "*ebb and flow*", a technique in which an automatic siphon or "bell siphon" is used to drain the water and which allows tides to be set up at

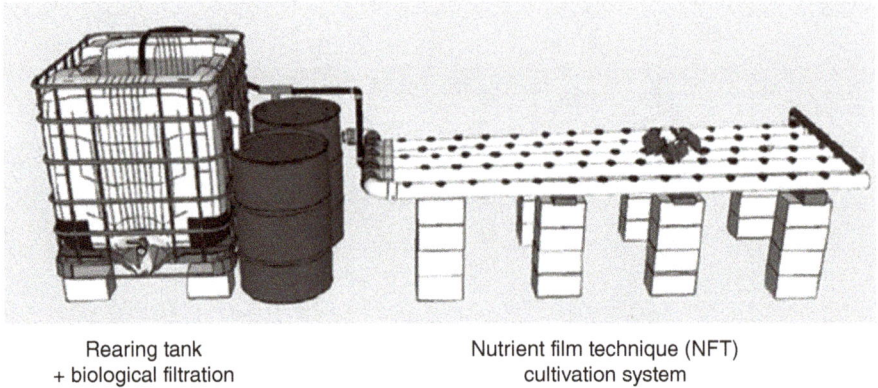

Rearing tank
+ biological filtration

Nutrient film technique (NFT)
cultivation system

Figure 2-16. Schematic diagram of the NFT culture principle. (Somerville *et al*., 2014)

Figure 2-17. Growing lettuce on NFT gutters. (Denfer007)

regular intervals, optimising oxygenation of the substrate. Figure 2-18 shows a diagram of this growing technique.

Provided that the volume of media is adequately sized in relation to the quantity of waste produced, dissolved and solid organic waste from fish farming can be directly decomposed and mineralised within this substrate by the action of heterotrophic bacteria, and earthworms if they are added beforehand: this is valid on a small scale with low production volumes and low fish farming densities. One of the shortcomings of this technique lies in the risk of deposition and accumulation of mineral elements (especially phosphorus and calcium) in the culture media over the long term, which considerably impairs plant nutrition by inhibiting the absorption of other essential nutrients. To eliminate these accumulations of salts, complete leaching of the rooting bed and growing medium is often necessary on a regular basis (Jones, 2005). In the short to medium term, in the case of a commercial installation, excessive accumulation of organic matter and clogging of the media can also be expected, leading to the release of compounds toxic to fish and plants, and/or a drastic drop in oxygen levels. Mechanical filtration with an efficient particle filter (such as a drum filter) upstream of the plant culture compartment is essential for any commercial system.

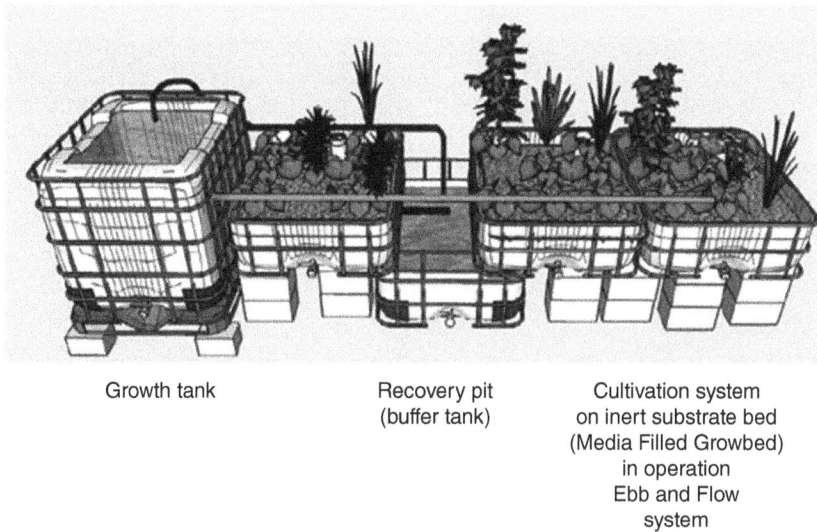

Growth tank   Recovery pit   Cultivation system
                     (buffer tank)   on inert substrate bed
                                           (Media Filled Growbed)
                                           in operation
                                           Ebb and Flow
                                           system

Figure 2-18. Schematic diagram of the MFG culture principle. (Somerville *et al.*, 2014)

11.3.4. Comparative approach to DWC, NFT and MFG methods

With the MFG technique, the substrates alone constitute the culture medium and are implanted in large quantities to fill the culture trays. With the other techniques, these substrates or media are contained in pots, which are themselves placed in the *rafts* or NFT gutters so that the bottom of the pot is just a few millimetres above the surface of the water, to ensure that the substrate remains moist at the start of cultivation; in time, the roots develop and end up soaking directly in the water as a result of gravitropism. Nutrients are then pumped into the water and not the substrate. Oxygen is also pumped directly into the water in dissolved form via the roots, and it is essential in all cases to ensure a sufficient level for plant growth, which can be done using bubblers or by simply moving the water (waterfalls, movements). In MFG, particular care must be taken to ensure that the substrates are not completely immersed in water: we generally ensure that a height of around 5 cm remains above water, while the rest is wetted by capillary action. The important thing is to maintain gas exchange zones at the top of the roots and at the neck of the plants. This also prevents the seeds from being displaced or drowned if they germinate directly in the media or substrate.

For *rafts* and NFT techniques, it is recommended that the seeds are first germinated on blocks of rockwool, peat or other substrates/media, which may or may not be soaked in a fertiliser solution for germination, or even directly in water from the aquaponics system, the important thing being to have the optimum atmospheric temperature for germination for a given species. Once a seedling stage has been reached, the clods are placed directly in pots and integrated into the growing media. It is not advisable to germinate seeds in a biologically active (non-inert) substrate when working above ground. Incorporating soil from outside could increase the risk of mould and other plant pathogens entering the system (*Pythium, Fusarium, Ascochyta, Phoma*, etc.). In MFG, it is possible to germinate seeds directly in the culture media, without any other substrate, because the roots will be able to settle to a sufficient depth to keep the plant in equilibrium.

Lennard and Leonard (2006) carried out a study to compare the yields achieved by these three types of system. Their conclusions were as follows: the yield efficiency of the MFG method was superior to that of the *rafting* technique, followed by the NFT technique. However, this study was carried out over 21 days, which seems insufficient to measure the long-term effects of the three systems. Maucieri *et al* (2017) also state, based on a meta-analysis of 122 publications, that the NFT technique is the least effective in terms of plant yield and water purification performance.

Existing commercial aquaponic productions to date mainly use the *raft* technique, which has many advantages and bearable disadvantages. Love *et al* (2014) conducted statistical studies on the use of these different techniques (alone or simultaneously) based on surveys of 'aquaponic farmers' around the world: 86% use inert substrates such as gravel or coconut fibre, 46% use the *raft* technique, 19% use the horizontal NFT technique, and 17% use vertical towers with root spraying.

Each of these techniques has advantages and disadvantages that need to be understood and taken into account when making the technical choices for an aquaponics project (table 2-5).

Table 2-5. Comparative study of the three main aquaponics techniques. (Pierre Foucard based on Rakocy *et al*, 2006; Connoly and Trebic, 2010; Lennard W., 2010)

Technical	Benefits	Disadvantages
Rafts	Large water storage volume (ideal for storing and treating fish farm effluent)	Filling the system with nutrient-rich water from the fish farm can take several weeks.
	Thermal inertia linked to the mass of water stored, useful in a greenhouse context	Sometimes difficult to manage humidity levels in the greenhouse
	Reduced contact between water and air and thermal insulation of the water mass	Need to add mechanical and biological filters
	High physico-chemical stability	High prevalence of inexpensive but environmentally unfriendly materials such as polystyrene in the design of culture plates
	Irrigation and oxygenation even over time	Request for interview
	Possibility of setting up *low-cost* structures	Limited range of plants suited to this technique (leafy plants, aromatic plants, short-cycle plants)
	Strong resilience to pump breakdowns	Need to add growing substrates (clods)
	High durability of culture plates	
NFT	High water efficiency, suitable for low fish/plant ratios and/or mixed aquaponics/hydroponics cultivation	Risk of clogging hydraulic circuits
	Practical and ergonomic	Low physico-chemical stability
	Easy maintenance	Low thermal inertia
	A wider range of plants adapted to this technique than rafts (strawberries, plants requiring a less humid environment than *rafts*)	Low resilience to pump failure
		Limited range of plants suited to this technique (leafy plants, aromatic plants, short-cycle plants)
		Expensive structures
		Need to add growing substrates (clods)
		Need to add mechanical and biological filters
Gravel bed	Acts as a mechanical and biological filter	Irrigation and heterogeneous oxygenation
	Media support plants	Risk of media clogging and physico-chemical imbalance
	High water efficiency, suitable for low fish/plant ratios and/or mixed aquaponics/hydroponics cultivation	Risk of anaerobic zones forming
	Broader range of plants suited to this technique than *rafts* (root vegetables, etc.)	Accumulation of minerals (calcium, phosphorus, etc.) that can affect the bioavailability of other nutrients
		Additional media costs (large volumes required, expensive)
		Not sufficient to treat solid and dissolved effluent from intensive RAS fish farming systems

11.3.5. The tidal table cultivation technique (or subirrigation)

It is a traditional technique in ornamental horti-culture, which makes it more transposable for a grower wishing to diversify his activity with aqua-ponics. This technique is quite similar to the MFG method, and corresponds in some ways to the 'professional' version of this growing system. The difference lies in the fact that this method is based on the use of horticultural shelves and 0.5 to 2 litre pots filled with inert substrates. Irrigation is controlled according to pre-established cycles and crop parameters: growth stage, substrate humid-ity and climatic factors. When the substrate is evenly moistened, water pumping stops and the nutrient solution is evacuated by gravity into the water retention tank. The substrates most com-monly used in subirrigation are peat and coconut fibre, often mixed in a 50%/50% ratio.

11.3.6. The drip irrigation technique (or percolation)

This technique is based on the use of drippers or capillaries to manage watering finely according to need. It is mainly used in horticulture and soilless market gardening. The nutrient solution is distributed directly at the base of the plant and percolates as close as possible to the root system. Surplus nutrients are evacuated through drain-age slits and returned to collection gutters, where they can be re-injected into the general circulation of the system. Professional soilless growers often use rock wool or coconut fibre 'loaves' because of their excellent water reten-tion capacity. Coconut fibre is a relatively new material that is starting to appear in production systems and is a good alternative to rock wool, as it is fully biodegradable and has good water re-tention and aeration properties.

The art of the percolation system lies in finding the right compromise between humidity and substrate aeration. To do this, you need to vary the duration and frequency of irrigation, bearing in mind that frequent short-duration ir-rigation is preferable to prolonged flooding of the substrate for root aeration. Schmautz *et al* (2016) report that the drip technique is more ef-fective than the *raft* and NFT techniques for growing tomatoes in terms of yield and the quantity of marketable tomatoes.

11.3.7. "Vertical" culture

Vertical cultivation makes it possible to optimise floor space, and in theory to achieve higher pro-duction than horizontal cultivation for a given surface area. It is an interesting approach for aquaponics in urban environments, particularly on the roofs of buildings, where the aim is gener-ally to use light production structures with low water content to avoid problems associated with the load-bearing capacity of the roofs. There are two ways of growing this type of crop. One is to superimpose several layers of horizontal cultiva-tion of the *raft* or NFT type, with LED lighting above each layer, given the shading that super-imposition implies (Figure 2-19). The aim is to be more profitable per square metre, despite the major investment in LEDs. This investment can be offset by their long life and low energy con-sumption. Another option is to use growing tow-ers in a vertical configuration, of the NFT type or other tower concepts containing a synthetic, highly aerated matrix that acts as a growing me-dium and allows roots to be irrigated with thin streams of water (Figure 2-20). These systems are more labour-intensive than *rafts* or NFT sys-tems (maintenance and handling). In addition, the towers need to be positioned in such a way as to limit the effects of the shadows cast by the towers in relation to each other.

11.4. Diagnosis and correction of mineral deficiencies in plants

Most of the information contained in this sec-tion is adapted from specialist works: Hopkins (2003) and Jones (2014), as well as a summary document from the FAO (2014).

11.4.1. General information on nutritional deficiencies in plants

Oxygen deficiencies can affect plants and can be confused with nutrient deficiencies, which is why it is important to monitor this parameter in the plant compartment: poor aeration of the root system can be observed in the crop, with roots colonising mainly the periphery of the substrate and tending to rise towards the sur-face, the most aerated area of the growing me-dium; leaves wither and turn yellow rapidly; a

Figure 2-19. *Rack* and *pinion* system at Green Spirit Farms in the U.S. (Élise Fargetton, ASTREDHOR)

loss of vigour and inhibition of growth appear and plants become more susceptible to root pathologies such as *Pythium*, *Fusarium* and *Verticilium*.

The symptoms of mineral deficiency depend in part on the mobility of the elements in the plant:

–   certain elements are immobilised once they are in the tissues and cannot be mobilised by the plant to form other tissues. In this case, the deficiencies are visible on the young leaves or stems. This is the case for calcium, iron, manganese, zinc, copper, sulphur and boron;
–   other elements can be mobilised by the plant to form young, developing tissues: in these conditions, deficiency symptoms appear primarily in the oldest tissues. This is the case for nitrogen, phosphorus, potassium and magnesium.

The pH has a major impact on the bioavailability of these macro- and micronutrients, even if they are present in sufficient quantities in the water. Since each element has one or more structural or functional roles, their absence or excess is manifested by the appearance of biochemical or morphological symptoms linked to this (or these) deficiency(s) or excess(es).

11.4.2. Macro-element deficiencies

There are eight macro-nutrients that plants need in relatively large quantities. These nutrients are nitrogen, phosphorus, potassium, calcium, magnesium, sulphur, carbon and oxygen. These last two elements are captured by plants from the air via their leaves (for carbon and oxygen, with O_2 and CO_2 gas) and from water via their roots (for oxygen, with dissolved $O_{(2)}$ gas).

11.4.2.1. NITROGEN (N). Nitrogen is the building block of proteins. As such, it is a major structural component of plants, used in particular for cell growth, metabolic processes and chlorophyll

Figure 2-20. ZipGrow vertical tower (Bright Agrotech)

production. As such, nitrogen is the most common mineral element in plant matter after carbon© and oxygen (0). Nitrogen is a key element in the hydro- or aquaponic nutrient solution. It is a highly mobile element in plants.

N deficiency: This is easily detected by the yellowing of old leaves (Figure 2-21), stunted stems, lack of vigour and reduced growth (Figure 2-22). Nitrogen can be reallocated in the plant tissues after mobilisation in the old leaves and lead to new growth. Nitrogen deficiencies in aquaponics are highly unlikely when the system is correctly dimensioned, as nitrates are produced in excess by the fish compared with other minerals. A nitrogen deficiency would therefore first imply a deficiency in all the other elements.

Excess N: this can lead to excessive vegetative growth, resulting in lush, dark-green plants that lack rigidity and brittleness, are more susceptible to disease and insect attack, and have difficulty flowering and/or forming fruit. Nitrogen is mainly provided by nitrates for good plant growth. Too much nitrogen in the form of ammonia also depletes the plant of carbohydrates and reduces growth, while lesions and/or rots can appear at the base of stems and on fruit

Figure 2-21. Evidence of nitrogen (N) deficiency on lettuce - chlorosis on older leaves first. (Mattson and Merrill, 2015)

Figure 2-22. Evidence of nitrogen (N) deficiency in lettuce, a stunted plant with stunted growth. (Mattson and Merrill, 2015)

in the case of fruiting plants. Magnesium deficiency symptoms can also be induced by excessively high concentrations of ammoniacal nitrogen.

11.4.2.2. PHOSPHORUS (P). Phosphorus is used by plants to form DNA molecules, but also as a structural component of the phospholipid membranes of cells, and of ATP molecules (Adenosine Tri-Phosphate, a cellular energy storage component). It is essential for the photosynthesis process, as well as for the formation of oils and sugars. It stimulates seed germination and root development. It is a highly mobile element in the plant.

P deficiency: this is often blamed for poor root development and very slow growth, and most often results in intense green coloration and a burnt appearance of the leaf tips (necrotic spots) as shown in Figure 2-23 (at a fairly early stage of detection). In some cases, it can be diagnosed by an accumulation of anthocyanin molecules in the oldest leaves, giving them a dark green to purple colour (Figure 2-24). P deficiencies in aquaponics are rare when the system has been optimised with a good aquaculture/plant compartment ratio. They can occur when the mechanical filtration technique is "too" efficient, thus removing too much organic matter from the system, and not allowing the fine solid

particles rich in phosphorus to be mineralised in the form of orthophosphate ions, in addition to those naturally present in the dissolved waste of the fish.

Excess P: this has the opposite effect to excess nitrogen. Root growth is stimulated in preference to leafy stem growth, thus reducing the leaf/root volume ratio. Excess phosphorus can also interfere with the assimilation of zinc, iron, manganese and calcium. This is unlikely to happen in aquaponics.

11.4.2.3. POTASSIUM (K). Potassium is used for cell signalling, a complex communication system that regulates the fundamental processes of cells and coordinates their activity, via controlled exchanges of ions across membranes. Among other things, potassium is involved in the opening of stomata and in flowering. It is also involved in the production and transport of sugars, water uptake, disease resistance and fruit ripening. It is a highly mobile element in the plant.

K deficiency: this manifests itself as burnt spots all over the surface of the oldest leaves (figure 2-25), and a lack of vigour and turgidity in the plant. Yellowing between the leaf veins (interveinal chlorosis) or drying of the leaf periphery may be observed. In some cases, the leaves may curl up and dry out (Figure 2-26).

Figure 2-23. Evidence of phosphorus (P) deficiency on lettuce - burnt appearance of leaf tips. (Mattson and Merrill, 2015)

Figure 2-25. Evidence of potassium (K) deficiency on lettuce - burnt-looking interveinal necrotic spots on older leaves first. (Mattson and Merrill, 2016)

Figure 2-24. Evidence of phosphorus (P) deficiency in basil - anthocyanin accumulation and purple coloration (Mattson and Merrill, 2016).

Without sufficient potassium, flowers and fruit may not develop properly. Potassium deficiencies can also be linked to excess calcium or magnesium in the solution. This type of deficiency is common in aquaponics, but can easily be remedied by adding potassium bicarbonate or carbonate to the fish farm water ($KHCO_3$ or K_2CO_3),

Figure 2-26. Evidence of potassium (K) deficiency on lettuce - older leaves first curl up and dry out (Mattson and Merrill, 2016).

which also acts as a pH buffer to manage the operation of the biological filter. Alternatively, foliar spraying twice a week with a 0.5 g/l potassium sulphate solution (Roosta, 2014a) is a more targeted and faster, but time-consuming, method.

Excess K: this can increase plant susceptibility to certain fungi responsible for root rot and lead to calcium and/or magnesium deficiencies in the solution. Such excesses are fairly unlikely in aquaponics, as this element is often the limiting element in the nutrient solution.

11.4.2.4. CALCIUM (CA).   Calcium is used as a constituent of cell walls and membranes. It is involved in strengthening the rigidity of stems and contributes to root development. It also plays an important role in the process of cell division linked to plant growth, and in regulating the activities of numerous enzymes. It has very low mobility within the plant.

Ca deficiency: this is fairly common in hydroponics and is always apparent in the youngest parts of the plant, since calcium does not circulate in the plant once it has been stored. Marginal necrosis of young salad leaves (Figure 2-27) or blossom end rot on tomatoes and courgettes are examples of symptoms. New leaves are often irregularly shaped, stunted or curled up, and fruit rots at the apical level. A clever way of treating these deficiencies is to use calcium bicarbonate $Ca(HCO_3)_2$, which acts as a pH buffer for the biological filter, just like $KHCO_3$. It is also possible to have calcium deficiencies even though calcium levels are high in the nutrient solution, particularly when plant evapotranspiration is very high (CTIFL, personal communication).

Excess Ca: this can make potassium and magnesium unavailable to plants, leading to the deficiency symptoms characteristic of these elements.

11.4.2.5. MAGNESIUM (MG).   Magnesium is a major component of the chlorophyll molecule, the main assimilating pigment in plants that enables photosynthesis. It also helps activate several plant enzymes required for growth and contributes to protein synthesis. It is a highly mobile element in plants.

Mg deficiency: this can be diagnosed by yellowing of the leaves with the veins remaining green, particularly in the oldest parts of the plant (Figure 2-28). Magnesium deficiency symptoms can be caused if there are too high levels of calcium, potassium or sodium in the substrate or nutrient solution.

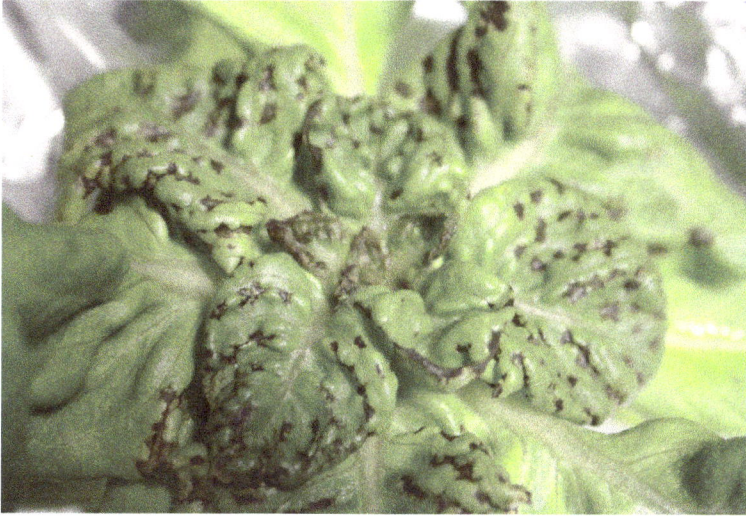

Figure 2-27. Evidence of calcium (Ca) deficiency in lettuce - irregular shape of new leaves, stunted appearance, necrosis, progressive death of apex (Mattson and Merrill, 2015).

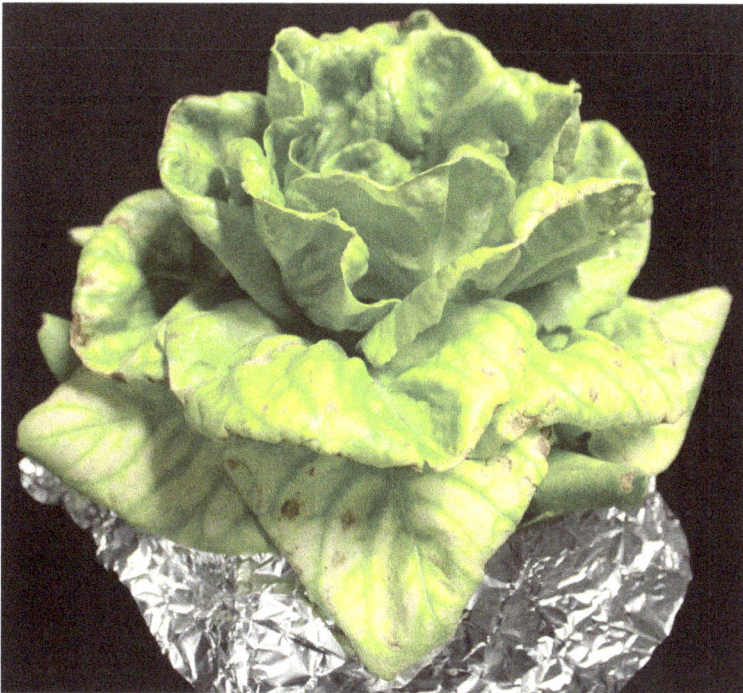

Figure 2-28. Evidence of magnesium (Mg) deficiency on lettuce - interveinal chlorosis and marginal necrosis on older leaves first. (Mattson and Merrill, 2015)

Excess Mg: this can make potassium and calcium unavailable to plants, leading to characteristic deficiency symptoms for these elements. New leaves may be mottled.

11.4.2.6. SULPHUR (S).   Sulphur is essential for the production of chlorophyll and certain enzymes governing the photosynthesis process. In addition, methionine and cysteine are two amino acids containing sulphur and which contribute to the formation of the tertiary structure of certain proteins. Sulphur is not a very mobile element in plants.

S deficiency: this is exceptional in both hydroponics and aquaponics, and results in widespread yellowing of young foliage. Leaves may also stiffen, become brittle or even fall off. This deficiency can resemble a nitrogen deficiency with widespread chlorosis and fragility of the plant, as shown in Figure 2-29 in a case that is already very advanced.

Excess S: leads to premature ageing and degradation of plants.

11.4.3. Micro-nutrient deficiencies

11.4.3.1. IRON (FE).   It is an element of major importance in plants, and is often limiting in aquaponics. It has a number of essential functions: on the one hand, it forms part of the catalytic group of numerous enzymes involved in oxidation-reduction reactions within the cells, and on the other, it is essential to the plant's respiration process and the synthesis of chlorophyll. It also helps to reduce nitrate and sulphate levels in the plant and to produce energy. It is not a very mobile element in the plant. Iron can be found in two distinct oxidation states:

– ferrous iron (Fe^{2+}) is a soluble form that is present at low pH and can be widely assimilated by plant roots;
– ferric iron (Fe^{3+}) is an insoluble form present at high pH, which is very difficult for plants to assimilate, appearing as a precipitate at pH 7 and above.

Fe deficiency: initially characterised by interveinal chlorosis due to a loss of chlorophyll, followed by widespread chlorosis all over the foliage and eventually by white leaves with necrotic spots and deformed leaf margins. Deficiencies occur more rapidly in young leaves (Figure 2-30) because iron is a non-mobile element in the plant. Iron deficiency is almost systematic in aquaponics. Iron can be supplied in chelated form - otherwise known as 'sequestered' iron. With this method, iron is released very slowly in a form that is bioavailable to plants. There are several types of chelated iron, including EDTA, DTPA and EDDHA. EDTA is considered toxic to aquatic animals and has the disadvantage of being poorly available below pH 7. DTPA is strongly recommended in the literature for use in aquaponics: it begins to precipitate above pH 7 and is ineffective above pH 8; EDDHA is also

Figure 2-29. Evidence of sulphur (S) deficiency on lettuce - stunted plants and chlorosis on young leaf first and rapidly on overall foliage. (Mattson and Merrill, 2015)

Figure 2-30. Evidence of iron (Fe) deficiency in basil, with interveinal chlorosis mainly on young leaves. (Pierre Foucard, ITAVI)

very interesting given that it only begins to precipitate above pH 10. The major problem with EDDHA is its unfortunate tendency to colour the water purple-red. So DTPA, which is also cheaper, seems to be the right compromise. There is also HBED chelated iron, which can work up to pH 12, with a less intense colouration of the water than with EDDHA: only its price remains prohibitive.

In aquaponics, 2 g of iron/m³ every two or three weeks is recommended (Rakocy *et al.*, 2006). The percentage of iron ion (x%) in the chelated iron is a variable to be considered. For example, to achieve 2 g iron/m³, based on x% EDDHA iron, you need to multiply 2 gFe/m³ by $1/x\%$, where x% represents the iron inclusion rate (typically between 4% and 12% depending

on the product). The value obtained must then be multiplied by the volume of water (m³) to obtain the mass (in grams) of product to be added to the water, approximately every 3 weeks. The weight can be converted into volume using the density indicated on the product data sheet.

Application methods: chelated iron can either be added directly to the water so that it can be taken up by the roots, or applied by foliar spraying, a more targeted method. It should be noted that while foliar application is ideal for a rapid response, it should also be borne in mind that iron is a nutrient that is not mobile within plant tissues: it will have to be applied regularly to feed young leaves, which makes it a time-consuming method.

Excess Fe: highly unlikely in aquaponics, this can lead to the appearance of black spots and brown discolouration on the leaves.

11.4.3.2. BORON (B). Boron (B) acts as a molecular catalyst, being involved in the formation of polysaccharides and structural glycoproteins, as well as in the transport of carbohydrates and the regulation of certain metabolic pathways in plants. It is also thought to be involved in reproduction and in the process of water absorption by cells. It is used with calcium in the synthesis of cell walls and is essential to the process of cell division. Boron requirements are much higher for the reproductive phase: it helps with pollination and the development of fruit and seeds. It is not a very mobile element in the plant.

B deficiency: this can be diagnosed by incomplete development of buds and flowers, shortening of internodes (Figure 2-31), giving the plant a bushy or rosette-like appearance with a thickened stem; brown spots can often be seen on the stems: the apex dies back and the lower shoots develop. The roots are yellow or brown, wrinkled and rotting at the base.

Excess B: causes browning and premature death of plants.

11.4.3.3. MANGANESE (MN). Manganese is used in plants as a major contributor to various biological systems, including photosynthesis, respiration and nitrogen assimilation. Manganese is also involved in pollen germination, pollen tube growth, root cell elongation and resistance to root diseases. Manganese is not very mobile in plants.

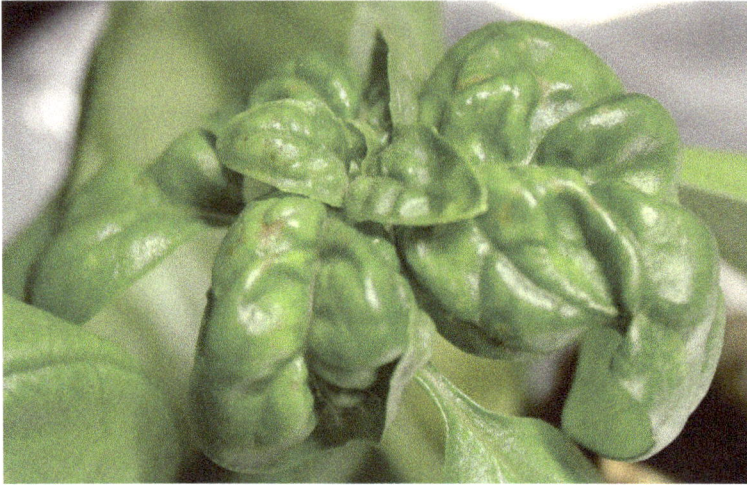

Figure 2-31. Evidence of boron (B) deficiency in basil - distorted appearance of young leaves and shortening of internodes (Mattson and Merrill, 2016).

Mn deficiency: this manifests itself as interveinal chlorosis of the leaves, progressing to brown necrotic spots. The veins remain green. Symptoms are similar to iron deficiency and include chlorosis. Manganese uptake is very low at pH above 8.

Excess Mn: this can be characterised by the appearance of brown spots on old leaves. It can also lead to iron deficiency by interfering with its uptake by the roots.

11.4.3.4. ZINC (ZN).   Zinc is used to activate numerous enzymes responsible for the synthesis of certain proteins. Its presence in plant tissue helps plants to withstand cold temperatures. Zinc is essential for the formation of auxins, which help regulate growth and stem elongation. It therefore has an impact on growth and ripening. It is not a very mobile element in the plant.

Zn deficiency: visible when plants are not very vigorous, with reduced internode length and leaf size; interveinal chlorosis is also present and can be confused with other deficiencies such as boron.

Excess Zn: plants can tolerate high concentrations of zinc (> 5 mg/l), but too high a concentration can interfere with iron uptake.

11.4.3.5. COPPER (CU).   Copper is used in the functioning of certain enzymes, particularly in reproduction. It helps to strengthen stems by stimulating lignin synthesis. It is also required in the process of photosynthesis and plant respiration, and helps in the metabolism of carbohydrates and proteins. Copper is also used to intensify the flavour and colour of vegetables and flowers. It is not a very mobile element in plants.

Cu deficiency: this can lead to interveinal chlorosis of young leaves, with leaf tips turning brown/orange. Fruits stop growing and become necrotic.

Excess Cu: this results in damaged roots if the level exceeds 0.3 mg/l.

11.4.3.6. MOLYBDENUM (MO).   Molybdenum is an essential component for the proper functioning of the two enzymes that convert nitrate into nitrite (a toxic form of nitrogen for plants) and then into ammonia (within the plant and not in the nutrient solution), which is then used to synthesise amino acids. It is also required by symbiotic bacteria that fix atmospheric nitrogen in legumes. Plants also use molybdenum to convert inorganic phosphorus into organic forms. It is a relatively mobile element in plants.

Mo deficiency: this can disrupt the development of flowering and fruiting plants. Symptoms such as chlorosis first appear on the oldest leaves. Necrosis may then appear.

Excess Mo: molybdenum toxicity is very rare, as plants are highly tolerant.

11.4.3.7. CHLORIDE (CL⁻) AND SODIUM (NA)⁺. Chloride and sodium are elements that are generally considered to be useless for plants, and even toxic at excessively high levels. However, research has shown that plants do need these elements in small quantities. Chloride is essential for photosynthetic reactions. It maintains electrical neutrality across membranes and helps maintain ionic balance in cells. Sodium is absorbed in the form of Na^+. It helps maintain cell turgidity. Sodium is also useful in small quantities for general metabolism and chlorophyll synthesis.

Excess Na or Cl: sodium toxicity results in necrosis or burning of leaf tips or margins. Chloride toxicity starts with premature yellowing of the leaves and then leads to necrosis of the tips or edges of older leaves.

11.4.3.8. CONCLUSION ON PLANT DEFICIENCIES. The aim of aquaponics is to do without synthetic fertilisers as much as possible, particularly as regards the elements nitrogen, phosphorus and potassium, which make up the bulk of fertilising elements. It is possible to supplement nutrient solutions in aquaponics if this proves necessary for certain types of plant and for elements in particular that can quickly become limiting, as is the case with iron. Ideally, micronutrient deficiencies (iron, boron, zinc, copper, molybdenum, manganese) should be anticipated by providing them fortnightly in the form of mixes formulated by specialist companies. These elements are often present in very small quantities, and providing them optimises plant growth and resilience to health problems. They can be added simply, at a lower cost, and with no environmental impact, especially when they are added in such a way as to precisely meet the needs of the plants, which will then extract them from the water.

Figures 2-32 and 2-33 can be used to help diagnose plant deficiencies so that appropriate and proportionate action can be taken. Magnesium and calcium are rarely limiting and can be added in small doses if necessary (in the form of magnesium sulphate or calcium sulphate). Nitrogen and phosphorus are not limiting factors as long as the sizing is optimised. Finally, potassium can be limiting depending on the type of fish feed and the quality of the water. It can be skilfully added by means of a pH buffer, often necessary in recirculated aquaculture to manage the pH of the water, which tends to fall over time due to the acidification of the water caused by nitrifying bacteria. The fact that aquaponics is totally self-sufficient in nitrogen and phosphorus is no mean feat, given that these two elements are the most worrying for the integrity of aquatic environments. No chemical fertilisers containing nitrogen or phosphorus are needed to grow plants in aquaponics when the system is properly designed.

11.5. Plant performance in aquaponics

The plant species most commonly grown in aquaponics systems are lettuces, which have been tested at various densities (from 20 to 40 plants/m²) and growing times (3 to 4 weeks in the optimum period, up to 7 to 8 weeks in winter in temperate regions). Yields are also variable, ranging from 1.4 to 6.5 kg/m² (Seawright *et al.*, 1998; Lennard and Leonard, 2006; Dediu *et al.*, 2012). Basil is also a model aquaponics crop, with densities of up to 40 plants/m² and yields ranging from 1.4 to 4.4 kg/m² for successive growing/harvest cycles averaging 28 days in warm regions (Rakocy *et al.*, 2004). Warm water crops are also very productive, particularly water bindweed, which yields around 33 to 37 kg/m² in 28 days of cultivation at a density of 100 plants/m² (Endut *et al.*, 2010), while okra could reach 2.5 to 2.8 kg/m² in less than three months at densities of 2.7 to 4 kg/m² respectively (Rakocy *et al.*, 2004). Figure 2-34 shows figures from a study by Savidov *et al.* (2014) on the annual yields of various aromatic herbs and leafy vegetables grown in aquaponics. Figure 2-35 shows yield figures taken from the work of APIVA®; the crops were grown over a period of time corresponding to the duration of the crop, in the most suitable season for each species according to climatic parameters. For ease of reading and to facilitate comparison between plant varieties, yields have been calculated on a monthly scale, in kg/m²/month.

Another interesting result observed in various research studies concerns the comparison of yields between hydroponics and aquaponics. Most of the time, the performance of aquaponics is more or less equal to that of hydroponics (Savidov *et al.*, 2005; Graber and Junge, 2009; Pantanella *et al.*, 2010; Thorarinsdottir *et al.*, 2015).

```
                              ┌─────────┐
                              │ Oldest  │
                              │ leaves  │
                              └─────────┘
```

┌─────────────────────────┐   ┌─────────────────────────┐
│ Symptoms appear ONLY   │   │ Symptoms first appear   │
│ on the leaves of the   │   │ on older leaves,   │
│ lower part of the plants │   │ THEN spread to   │
│   │   │ the entire plant.   │
└─────────────────────────┘   └─────────────────────────┘

| The leaves appear pale green/yellow with interveinal chlorosis | Old leaves are very pale green, yellowing and deformed | The plant shows general yellowing with browning and drying of the leaves on the lower parts | The plant has a stunted appearance, with a dark green or even dark reddish-purple colour on the oldest leaves |

Magnesium   Molybdenum   Nitrogen   Phosphorus

The oldest leaves wilt and/or turn brown. Necrotic spots appear around the edges.
The distance between the nodes on the stem becomes shorter and the plant loses rigidity

Potassium

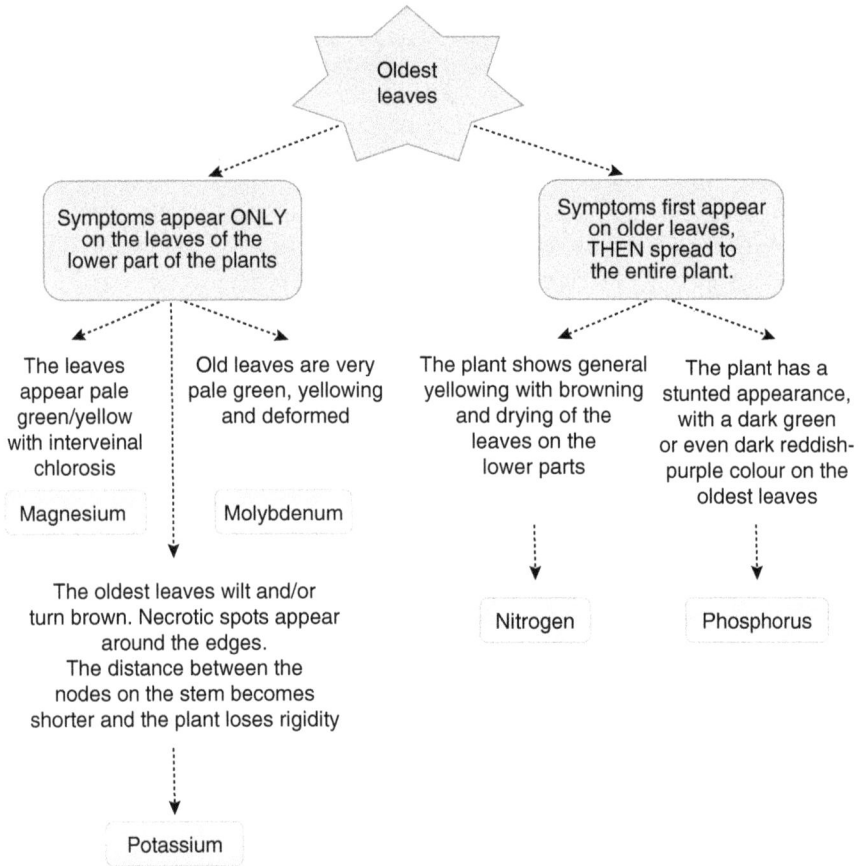

Figure 2-32. Plant deficiency determination decision grid for symptoms observed on older leaves as a priority. (Pierre Foucard adapted from Loper, 2014)

Table 2-6 shows some of the results obtained on this subject by these authors.

For leafy vegetables, there were no differences between aquaponics and hydroponics, in terms of leaf productivity and quality, chlorophyll quantity, leaf area and nitrate concentration in the leaves (Pantanella *et al.*, 2010, 2011a, 2011b, 2012). Rebalancing the nitrogen/potassium ratio (by adding potassium to the water in aquaponics) could lead to improved yields in aquaponics. Delaide *et al* (2016) reaffirmed this finding, while adding that an aquaponic fertiliser solution supplemented with hydroponic fertilisers - so as to meet the concentration standards for the various elements - increased lettuce yield by 39%, which raises questions about the very nature of the aquaponic fertiliser solution and its accompanying bacterial microflora, potentially beneficial for the plants.

11.6. Biological control in aquaponics

Aquaponics involves a wide biodiversity of micro-organisms, in the same way as in natural ecosystems. The bacterial flora present in aquaponics and the interaction of this flora with the environment are only emerging areas of research. Studies have shown that plants grown in aquaponic systems may be more resistant to the diseases that affect soilless and conventional crops (Gravel *et al.*, 2015). This resistance could be due to the presence of a stable and ecologically

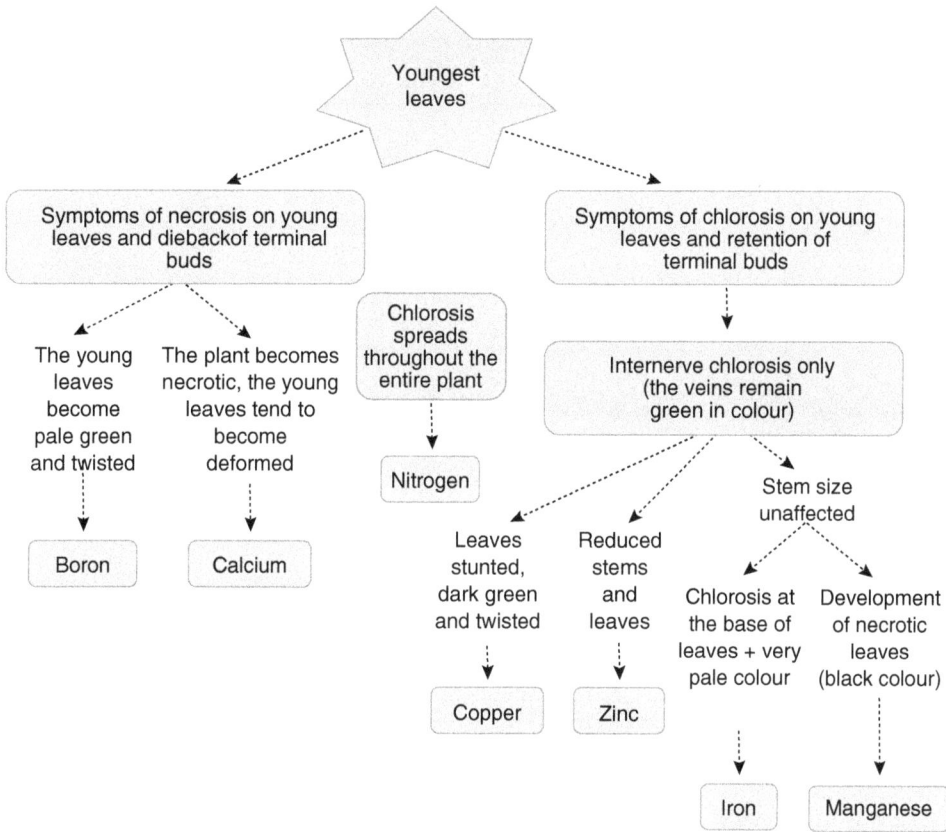

Figure 2-33. Plant deficiency determination decision grid for symptoms observed on young leaves as a priority. (Pierre Foucard adapted from Loper, 2014)

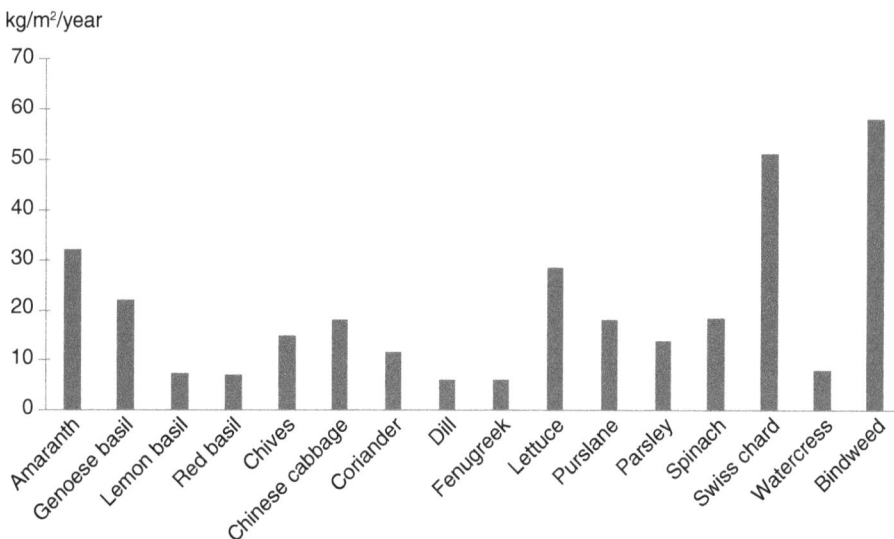

Figure 2-34. Yields of different plants (kg/m²/year) grown on *rafts* (Thorarinsdottir *et al.*, 2015 in Savidov *et al.*, 2010).

kg/m²/month

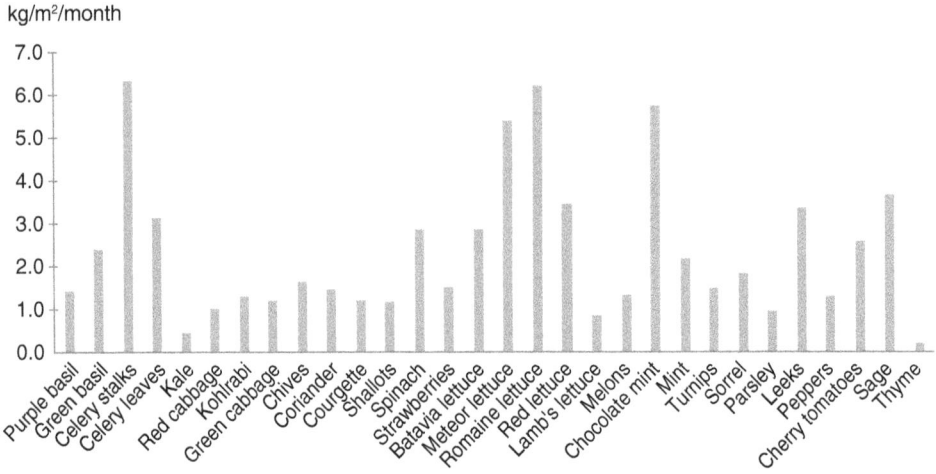

Figure 2-35. Yields of different plants (kg/m²/month) grown on *rafts*, and strawberries grown on coconut coir, measured as part of APIVA's® work. (ITAVI)

Table 2-6. Comparative yields obtained in aquaponics with a hydroponic control. (ITAVI based on Savidov *et al.*, 2005; Graber and Junge, 2009; Pantanella *et al*, 2010; Thorarinsdottir *et al.*, 2015)

Plant	Yield (kg/m²)		Source
	Aquaponics	Hydroponics	
Tomato	31-59	41-45	Savidov *et al.*, 2005
	23.7	26.3	Graber *et al.*, 2009
Cucumber	42-80	50	Savidov *et al.*, 2005
	3.3	5.2	Graber *et al.*, 2009
Aubergine	7.7	8	Graber *et al.*, 2009
Lettuce	5.7	6	Pantanella *et al.*, 2010

balanced environment, with a wide diversity of micro-organisms, some of which are thought to be antagonistic to the pathogens that affect plant roots (Rakocy *et al.*, 2006) and some of which could help plants to grow while forming a barrier against pathogens ('PGPR' bacteria, mycorrhizae, etc.). This microflora could have significant beneficial effects on plant growth and the assimilation of organic molecules (Goddek *et al.*, 2015), which would explain why yields are similar to hydroponics (and higher than in the open field), despite low concentrations of mineral elements.

Plants grown in aquaponics can also be affected by crop pests or diseases spread by pathogenic micro-organisms, as in any production system. The philosophy of aquaponics is to

completely dispense with the use of phytosanitary products, particularly because of the communication (in coupled systems) between the fish and hydroponic compartments and the impact of chemical molecules on the fish. This problem can be circumvented by decoupling the two compartments, a concept in which the plant water recirculates on itself and does not return to the fish part, while the daily volume of water leaving the fish system (in compensation for the volume of new water brought in and defined by the system's opening rate) renews part of the volume of water circulating in the hydroponic part on a daily basis. With sustainable development in mind, the aim is to limit phytosanitary treatments to extreme cases where no method of integrated biological protection would be sufficiently effective, especially as a "pesticide-free" label is a selling point that can play a role in consumer choice.

There are various integrated biological control alternatives:

- resistant cultivars;
- crop auxiliaries or predatory insects. These include ladybirds (*Adalia bipunctata*), chrysopid larvae (*Chrysoperla carnea*) and parasitic wasps such as *Aphidius ervi* for aphid control and *Encarsia formosa* for whitefly control (Fujiwara *et al.*, 2013). There are also predatory mites such as *Amblyseius swirskii* for whitefly control;
- antagonistic bacteria, such as *Bacillus thuringiensis* used to combat caterpillars;
- parasitic fungi such as *Beauveria bassiana*, to control whiteflies, aphids and thrips (Savidov, 2005; Blidariu, 2011; Tavares *et al.*, 2015).

In addition to these sometimes costly biological control methods, it makes sense to use preventive trapping and counting techniques (brightly coloured sticky strips) before the infestation spreads. There are also natural methods for treating insect colonies that have not yet developed: spraying black soap, which has antibacterial and insecticidal properties against aphids and is capable of killing larvae while cleaning up the honeydew that engulfs plant leaves. It is also effective to some extent against mealybugs, whiteflies, psyllids, thrips, etc. You can also add bicarbonate of soda to black soap to combat mildew, although the results will depend on how advanced the infestation is.

It is worth incorporating companion plants, such as 'trap plants', which attract pests by diverting them away from plants of commercial interest, or intercropping 'repellent plants', which scare off specific predatory insects because of the odours or pheromones they emit. The combination of the two is even more effective, and is known as the *'push-pull'* strategy described by Cook *et al* (2007), which consists of making the crop repellent to pests (*push*) while attracting them (*pull*) to areas where they can be managed (physical or chemical destruction), trapped or simply diverted from the crop at the susceptible stage. Flowering or nectar-bearing plants are also of interest as they provide a refuge and nutritional source for pest-predatory

crop auxiliaries that can come naturally or be artificially integrated into glasshouses as part of an integrated biological protection plan (Tavares *et al.*, 2015).

Biological control is an interesting alternative to chemical control. It can be applied in label production and as a component of integrated or integrated crop protection strategies. Bittsanszky *et al* (2016) and Goddek *et al* (2015) point out that very few tools are available for plant protection in aquaponics, and that the emphasis should be on precautionary measures to minimise infiltration by pests and pathogens (Junge *et al*, 2017), as well as good hygiene practices (Fox *et al*, 2012).

12. The bacterial compartment: micro-organisms

Microbial communities (bacteria, fungi and protozoa) located in the biofilter and on plant roots play a major role in nutrient dynamics in aquaponic systems (Bittsánszky *et al.*, 2016; Munguia-Fragozo *et al.*, 2015; Bartelme *et al.*, 2018). In addition to the nitrification activity of certain bacteria, others participate in the degradation of fine particles, leading in particular to the mineralisation of phosphorus. Both functions are absolutely crucial for the stability of an aquaponic system (Bittsánszky *et al.*, 2016; Munguia-Fragozo *et al.*, 2015). The bacterial component is currently a black box, and little is known about the nutrient requirements and potential sensitivities of these microbial communities to water quality (Kantartzi *et al.*, 2006; Bittsánszky *et al.*, 2016). The bacteria that play a role in aquaponics and/or recirculating fish farming systems can be grouped according to their functions.

12.1. Nitrification of ammoniacal nitrogen

Nitrifying bacteria are said to be "chemoautotrophic", which means that they can generate their own organic matter from mineral elements, using energy produced by chemosynthesis. These bacteria derive their energy from the oxidation reaction of ammoniacal nitrogen released by fish, which is a toxic form of nitrogen. This reaction leads to the formation of nitrates, a harmless form of nitrogen in high concentrations. This reaction is part of the nitrogen cycle, and forms the basis on which conventional and aquaponic recirculation systems operate. Bacteria of this type are found mainly in the biofilter, attached to the media provided for this purpose, but also in the plant culture substrate.

12.2. The mineralisation of organic matter

Heterotrophic" bacteria are capable of synthesising their organic matter from organic sources (solid fish waste, uneaten food particles). They can be found everywhere in the system, whether in the biofilter competing with nitrifying bacteria for oxygen, or in the rearing tanks and in the culture water. This mode of nutrition is characteristic of all living organisms that are neither chemoautotrophic (in the case of nitrifying bacteria) nor photoautotrophic (in the case of plants and cyanobacteria). These micro-organisms consume oxygen, so it is important to oversize oxygenation systems to take account of fish and bacteria, not only those in the biofilter but also heterotrophic bacteria. The bacterial compartment accounts for 15 to 30% of total oxygen consumption in a recirculated circuit.

These heterotrophic bacteria can be used skilfully to degrade and mineralise the sludge from the farm, in an aerated mineralisation tank separate from the system. In all cases, however, the aim should be to filter the farm water mechanically in order to remove as many solid particles as possible from the system. These are real ecological niches for heterotrophic bacteria, which can become a problem in excessive quantities, due to the consumption of oxygen and the release of ammoniacal nitrogen that they entail.

In the plant compartment, these bacteria will play a useful role in degrading the finest organic particles, which constitute a veritable reservoir of nutrients for plants. Da Silva Cerozi and Fitzsimmons (2016a) showed that inoculation of *Bacillus* spp. into the nutrient solution of lettuce grown in aquaponics resulted in better growth and higher levels of phosphorus and chlorophyll than in a control system: the ability of *Bacillus* to mineralise phosphorus being one of the explanations put forward.

12.3. Denitrification

Widely used in wastewater treatment, this inexpensive system is based on a biological phenomenon that takes place in anaerobic conditions under the effect of bacteria capable of reducing the NO_3^- ion to satisfy their oxygen requirements, leading to the formation of nitrogen gas (N_2). To do this, they need a simple source of organic carbon, such as ethanol. There is little data on the N_2/N_2O ratio at the outlet of these denitrification filters: nitrogen oxide N_2O is a well-known greenhouse gas. If anaerobic zones develop in a fish farming system, other bacteria can develop and cause denitrification, for example (transformation of nitrates into atmospheric nitrogen). The aim is to avoid this phenomenon in aquaponics, while commercial-scale RAS sometimes make use of this phenomenon to avoid the accumulation of nitrates in the rearing water.

12.4. Protecting and stimulating plant growth

Historically, research into aquaponics has been carried out by communities of researchers linked to aquaculture. Today, plant research is beginning to take an interest in the subject, and more specifically in understanding the interactions between plants and bacteria. It is interesting to consider the factors that explain why aquaponics is able to achieve good plant yields with water quality that does not meet the criteria for hydroponics as defined by decades of research and development: high levels of mineral nutrients, high conductivity, acid pH, ratios to be respected between certain minerals that are antagonistic to the root absorption process, etc.

These bacteria are known as PGPR (*Plant Growth Promoting Rhiziobacteria) and* are capable of stimulating plant growth. Various studies have documented improvements in the health and productivity of different plant species through the application of rhizobacteria that promote root growth and the bioavailability of nutrients in the soil (Ahemad and Kibret, 2013; Selosse *et al.*, 2004; Zaidi *et al.*, 2009; Ibiene *et al.*, 2012; Sharma *et al.*, 2013; Bartelme *et al.*, 2018). These PGPR rhizobacteria can reduce overall dependence on chemical inputs, which tend to destabilise agroecosystems (Bartelme *et al.*,

2018). Some PGPR are symbiotic (*Rhizobium, Bradyrhizobium, Mesorhizobium*) and others non-symbiotic (*Pseudomonas, Bacillus, Klebsiella, Azotobacter, Azospirillum, Azomonas*) (Ahemad and Kibret, 2014; Bartelme *et al.*, 2018). Some plants, such as legumes, also meet their nitrogen requirements through associations with prokaryotic symbiotic bacteria, the *Rhizobiaceae*, which have the ability to fix atmospheric nitrogen (Selosse *et al.*, 2004). In soilless environments, research on PGPRs is limited, but existing studies suggest that they play an important role in plant growth and health (Gravel *et al.*, 2006; Gravel *et al.*, 2015; da Silva Cerozi and Fitzsimmons, 2016a; Sheridan *et al.*, 2016). Mangmang *et al* (2015) experimented with the addition of *Azospirillum brasilense* to the seeds of various plants grown in aquaponics, and the results speak for themselves: an increase in root length and leaf area, higher levels of dry matter, protein and chlorophyll, and so on. *Azospirillum* is thought to have a direct effect on the plant, facilitating nitrogen fixation, the synthesis of phyto-hormones (in particular indole-3-acetic acid, IAA) and modulation of the plant's hormonal balance.

Other micro-organisms include mycorrhizal fungi. These micro-organisms, present in the soils of most ecosystems, form symbiotic associations with the roots of 80% of terrestrial plant species. They improve plant nutrition (especially in terms of less mobile elements such as phosphorus) and provide protection against pathogens and tolerance to heavy metals and organic pollutants (Sanon, 2005). In practical terms, a mycorrhiza is a zone of exchange: the plant receives nutrients collected in the soil and mineralised by the fungal partner, while the host plants provide a carbon exudate in the form of carbohydrates in exchange. In addition, positive interactions have been demonstrated between mycorrhizal fungi and soil bacterial communities (Sanon, 2005). The benefits of mycorrhizae for plants can be of an immune nature, providing enhanced protection against certain pathogenic micro-organisms: for example, the fungus *Trichoderma* (Gravel *et al.*, 2015) or arbuscular mycorrhizal fungi (Utkhede, 2006), which are

said to have a protective effect on plants against the phytopathogenic moulds *Fusarium* or *Pythium* in aquaponics and also in hydroponics (Aerts *et al.*, 2002; Rojo *et al.*, 2007; El Komy *et al.*, 2015), although the mechanisms involved are not fully understood: inhibition of the nutrition of pathogenic intruders or physiological changes in plants.

Mycorrhizae and rhizobacteria are just the tip of the iceberg of associations between plants and other living organisms. Their diversity is necessary in the field, because this complexity is part of well-established ecosystemic processes that lead to virtuous cycles of nutrient recycling and preservation of soil structure and composition. In conventional agriculture, the addition of phytosanitary products (fungicides, pesticides, etc.) and deep ploughing practices have harmful effects on soil quality by leading to a decline in the quantity and quality of fungal plant symbionts (Varma, 2008), as well as insect and worm populations (Ernst *et al.*, 2009), which play an important role in soil fertility and tillage.

This biodiversity, which is supposedly less present in soilless cultivation, is a criticism often levelled at this agricultural practice, despite the far superior growth performance of hydroponics. Aquaponics offers a new perspective on soilless cultivation, with the bacterial flora it provides leading to excellent plant growth performance despite the apparent poverty of nutrient solutions compared with the benchmark set by hydroponics. The first elements have recently appeared in the scientific literature on the quantification and determination of micro-organism populations in aquaponics. Schmautz *et al* (2017) carried out initial work on this topic using metagenomics, analysing the bacteria present in different compartments of aquaponic systems (samples of fish excrement, biofilm samples present in fish tanks and on biofilter media, and plant root samples). An interesting proportion of *Pseudomonas* was detected specifically in plant roots. This PGPR bacterium is useful in biocontrol, while certain rhizobacteria were also detected (*Acidovorax, Sphnigobium*...) without their role being fully defined (Sirakov *et al.*, 2016).

3

Designing and monitoring an aquaponics system

While the first chapter defined the principle of aquaponics and chapter 2 provided the basis for the operation and design of the three compartments inherent in any aquaponic system, the aim of this third chapter is to go into more detail on various technical points and to describe the basis for designing aquaponic systems, organ by organ. As water is the element common to each aquaponics compartment, the importance of monitoring its physico-chemical parameters will be explained and recommendations for compromises to be adopted to satisfy each organism will be given.

13. Approaches to sizing aquaponic systems

13.1. The benefits of a dimensioning approach upstream of a project

While an understanding of recirculating fish farming is necessary for aquaponics to develop, a good understanding of hydroponics is also essential. In aquaponics, the fish are fed a complete feed formulated according to their specific needs, and incorporate some of the nutrients before producing solid and dissolved waste: certain waste products are converted and/or made available by bacteria (autotrophic and heterotrophic) into nutrients that the plants can then

absorb and incorporate. This means that nutrient resources are shared between the initial source (feed), the initial user (fish), the intermediate converter (bacteria) and the tertiary user (plants).

An often overlooked facet of this integrated system is that commercially available aquaculture feeds are formulated to meet the exact nutrient requirements of the fish and not the requirements of the plants. As a result, the nutrient ratios and concentrations required for optimal plant growth and production are generally not present.

The main nutrient in commercial fish feed is nitrogen (N), as it contains a relatively high proportion of protein. Fish metabolise this protein nitrogen to produce amino acids for their own bodies, but much of this nitrogen is released in the form of ammonia. Another element that is very present in waste is phosphorus (P), which ends up partly in the form of solid waste (suspended matter) and partly in dissolved form (orthophosphates). Leafy plants such as lettuce generally require an N:P concentration ratio of between 3:1 and 5:1. Protein-rich feeds used to grow carnivorous fish such as salmonids or sturgeons generally involve a ratio of between 3:1 and 6:1, according to APIVA's® experiments, which is fairly consistent with this ratio. Less high-protein feeds used for breeding cyprinids, for example, give ratios of around 7:1 to 9:1. In the literature, ratios of between 3:1 and 30:1 are found for various species of fish.

©2026 CAB International. *Aquaponics* (eds Pierre Foucard and Aurélien Tocqueville)
DOI: 10.1079/9781836991441.0003

This shows the wide range of possible scenarios. Depending on the type of feed, the species of fish and the stage of growth, the nutrient discharges (N, P, K, Mg, Ca, S and microelements) will be quite variable, which makes it difficult to establish a typical fish/plant ratio applicable to all aquaponic systems, all the more so as the plants (species, growth stage, crop density) and the climatic (light, air temperature, air humidity) and physico-chemical (pH, conductivity, water temperature, redox potential, oxygen levels, etc.) parameters will also have an impact on consumption.) will also have an impact on the consumption of these nutrients by plants downstream of the fish farm.

13.2. A "fish/plant ratio" approach to sizing

The best-known and most studied aquaponics system, developed at the University of the Virgin Islands (UVI) by James Rakocy in 1980, has a ratio of plant culture surface area to aquaculture rearing surface area of 7.3:1, with a fish rearing density of 60 to 100 kg/m^3 and a feed based on pellets formulated to achieve a protein content of 30%. At the same time, the culture system is of the *rafts* type, where lettuce, basil and other leafy plants are grown in water with optimum physico-chemical characteristics for all compartments. The opening rate is less than 2%.

Other studies (Lam *et al.*, 2015) refer to a ratio of the volume of water in the plant compartment to the volume of water in the aquaculture compartment, with the 3:1 ratio offering the best compromise for plant growth: as the fish culture density is not specified, this type of data does not really provide any relevant information for sizing an aquaponics system.

Maucieri *et al* (2017), based on a review of 11 publications, found a ratio of plant culture area to water volume in the aquaculture compartment ranging from 1.2:1 to 8.7:1 for the "*raft*" method without being able to specify which is more suitable.

In short, there is no such thing as "fish/plant ratios", either in terms of surface area occupied or volumes of water used, which could be generalised for the sake of simplicity, so varied are the scenarios and system designs. Each configuration is unique, and that's what makes

aquaponics so complex. It is clear that nutrient levels will vary in each aquaponics system depending on the quality of the renewal water, the fish farming density, the feed used (ration plan, composition) and therefore the fish species, but also the type of plant grown and their phenological stage (Rakocy *et al.*, 2006; Timmons and Ebeling, 2007). It therefore makes no sense to aim for surface/surface, surface/volume or volume/volume ratios. We therefore need to start our reasoning from a different point of departure: the feed distributed to the fish is the source of nutrients for both the fish and the plants, so it is the quantity of feed that we need to take as the starting point for reasoning about the size of the planted area downstream.

13.3. Sizing approach based on the ratio of feed quantity to plant surface area

The most relevant sizing ratio was certainly the approach also developed by the University of the Virgin Islands, with the ratio of daily quantity of feed to plant surface area (which we will call "RASV" hereafter) in the case of the UVI system (Rakocy *et al.*, 2006).

13.3.1. Approach to the concept

The advantage of this ratio is that it eliminates certain variables linked to stocking density and fish rationing rates. The challenge is therefore to determine the quantity of fish feed needed to achieve a sufficient quantity of nutrients in the water to satisfy the requirements of a given biomass of plants on a given surface area. The RASV is defined as follows:

$$RASV = \frac{\text{Quantité d'aliment distribuée journalièrement aux poissons [g / j]}}{\text{Surface utilisée pour la croissance des plantes [m}^2\text{]}}$$

According to this work, the 'ideal' value for RASV varies between 60 and 100 g/m^2/day, based on empirical approaches and experience acquired over more than ten years of research: closer to 60 g/m^2/day for aromatic herbs such as basil, chives and parsley, or for leafy vegetables such as lettuce, and closer to 100 g/m²/day for vegetables and other plants with higher requirements, such as tomatoes and peppers (Rakocy et al., 2006). These ranges were determined empirically for a tilapia farming system and a lettuce and basil *rafting* system.

13.3.2. RASV calculation method

Based on the area available for cultivation, the number of plants that can be grown is determined according to the species chosen and its maximum cultivation density. The length of the production cycle is used to determine the yield per unit area and per species. We also estimate the appropriate daily quantity of fish feed to meet the requirements of the plants using the RASV, for a crop at maximum plant density. Based on this amount of feed, an appropriate fish biomass is determined according to the characteristics of the species, the growth phase (fry, juvenile, adult), the conversion index and/or the ration rate to be applied. Finally, the volume of water required to ensure good rearing conditions can be easily assessed on the basis of the fish biomass determined, by setting a maximum rearing density based on the technical characteristics of the installation (filtration and oxygenation performance). The aim of this approach is to optimise the sizing of rearing tanks based on a plant production objective. It is also possible to follow the opposite approach, starting from a desired fish production capacity, to determine a suitable surface area for plant cultivation.

13.3.3. The limits of the RASV approach

The RASVs determined in the scientific literature vary: Hafedh *et al.* (2008) showed that their UVI-type commercial system, with a hydroponic *raft* system, staggered lettuce crops and multiple tilapia rearing units, had an optimum RASV of 56 g/m²/day, with no plant nutrient deficiencies or excessive nutrient accumulation, over a period of one year. Endut *et al* (2010) found an optimum RASV of 15 to 42 g/m²/day for spinach grown on a bed of inert media. Other experts recommend a ratio of 40 to 80 g/m²/day (Somerville *et al.*, 2014).

These differences are not surprising, as comparing RASVs between different studies only makes sense if you are working with a fixed opening rate (ratio of volume of new water/ quantity of feed) and a given culture technique, with a given type of feed for a given species of fish and plant. Each species of fish has very specific nutritional requirements, which means that the composition of the feed will vary depending on the species, and the waste will also vary depending on their metabolism. The same can be said for plants, although it is possible to simplify the problem by classifying different plants into broad categories (leafy plants, fruiting plants) that have distinct needs. Only modelling nitrogen and phosphate discharges can really enable accurate sizing, by determining upstream a quantity of solid and dissolved effluents discharged for a given species of fish: these discharges will determine the quantity of nutrients remaining for the plants. INRA has developed a mathematical model for calculating discharges (NH_4^+, total phosphorus, suspended solids) for trout (Papatryphon *et al.*, 2005).

A second element that probably needs to be taken into account in the design (and which is not mentioned or taken into account in the Rakocy ratio) is the quantity of water evaporated in the system by the action of the plants. Depending on the type of plant, evapotranspiration will not be the same for the same surface area: water renewal and therefore the opening rate will be modified accordingly.

The relative proportions of each soluble nutrient from the aquaculture compartment do not always reflect the proportions that can be assimilated by growing plants. This can quickly lead to nutrient imbalances in the culture water, and to non-optimal ratios between certain antagonistic nutrients in the root uptake process (Seawright, 1998; Endut *et al.*, 2010). In addition, the species of fish used can have a major impact on plant growth, due to differences in digestibility for a given feed.

If such a ratio is used, it is important to ensure that the quantity of feed to be distributed changes as little as possible over time, so as not to unbalance the system. The choice and management of aquaculture and plant production strategies over time are therefore crucial. A study by Liang *et al* (2013) showed the benefits of increasing the frequency of feeding in order to achieve higher joint growth of fish and plants, while allowing the daily ration to be distributed more evenly over time, thus avoiding 'spike effects' and saturation of the plants' filtering capacity.

The RASV range calculated by Rakocy is therefore not a formula that can be applied "to the letter", but rather an indicative order of magnitude that needs to be corrected with experience and adapted according to a number of parameters: plant species cultivated and culture system, quality and composition of the aquaculture feed, particularly with regard to protein content, concentration of nutrients in the farm effluent (after extraction of TSS and/or biodigestion of sludge) and the opening rate of the system.

A new line of research could be developed to formulate feeds that offer a compromise between data on digestibility and use of nutrients by fish and data on assimilation of nutrients by target plants (Martins *et al.*, 2010; Endut *et al.*, 2010) in order to re-establish adequate ratios between the different elements. Theoretically, the nutritional composition of a feed can be modified so that the relative proportions of nutrients excreted by fish are closer to the relative proportions of nutrients assimilated by plants, as long as this does not impact on zootechnical performance. With such a diet, it would be possible to have a more efficient fish/plant ratio and optimal nutrient concentrations could be maintained for long periods without too much change in the system (Seawright, 1998; Endut *et al.*, 2011). This type of feed should ideally be enriched with micro-nutrients and potassium.

13.4. Mass balance approach to dimensioning

An alternative to the RASV method is the "mass balance" approach, which shows exactly how to balance the nutrient requirements of plants with fish farming discharges. The approach consists of quantifying (volume for fluids, mass for solids) and analysing (chemical composition in mg/l for fluids or in g/kg for solids) all the inputs and outputs of the system, over a given period of time, for example the duration of a plant production cycle, in order to draw up a balance sheet of nutrient flows in the system.

Inputs are generally the fish feed, the mass of water present in the system at the start of the balance, and the mass of renewal water added throughout the balance. Other inputs include products used to buffer the pH, such as potassium bicarbonate, as well as microelements that can be added as supplements, in particular iron.

The outputs are made up of the biomass of fish and plants produced over the duration of the balance, but also the biomass of sludge extracted from the system, the mass of water present in the system at the end of the balance, and the mass of water discharged by the overflow of the system throughout the balance (discharges dependent on the opening rate of the system).

To sum up, the mass balance consists of assessing the fraction of nutrients (N, P, K, etc., contained in inputs) retained by fish (for a given species) and by plants (for a given species or for a mixture of several species) during a production phase, as well as the fractions emitted in faeces (sludge) and those present in solution in the water (dissolved minerals and suspended matter). For nitrogen, the balance is often not balanced, unlike elements such as phosphorus and potassium, which are described as 'conservative'. The error in assessing the nitrogen balance is due to the fact that gaseous forms can escape from the system in various forms (NH_3, N_2O, N_2).

This makes it possible to assess the fate of each nutrient, to detect potential deficiencies for the plants, and to best estimate the appropriate sizing of the system so that fish and plants can cohabit harmoniously and without imbalance. These assessments are carried out for a given species of fish, with a given type of feed, for a species of plant or for several species at the same time, all in a given climatic and physico-chemical environment. In general, these approaches are carried out earlier in the design phase, in order to determine more precisely the design to be aimed for.

This makes it possible to determine precisely how much feed to provide to avoid phosphorus deficiencies. Nitrogen is rarely a problem, as this element is present in excess in aquaponics. Magnesium, potassium, calcium and sulphur can easily be supplemented as required, even with biological solutions that are harmless to the fish. In practice, only potassium generally needs to be added to supplement nutrient solutions in aquaponics.

A major advantage of aquaponics is that nitrogen and phosphorus can be supplied entirely by fish waste, and no longer need to be supplied by mineral fertilisers of chemical or mining origin. These elements are the most harmful to the environment, not only through their manufacturing or extraction *process*, but also through the eutrophication of aquatic environments that they cause.

The mass balance approach also ensures that all the benefits attributed to aquaponics are actually present:

–   efficient and optimised use of the nutrients released by the fish farm;
–   efficient and optimised use of water ;
–   maximised elimination of nutrient-rich waste streams, thereby minimising the impact on the environment.

Mass balances for aquaponics have so far been little studied. They are often carried out on a small scale (Seawright *et al.*, 1998; Endut *et al.*, 2010; Da Silva Cerozi and Fitzsimmons, 2016c; Delaide *et al.*, 2017), or take few nutrients into account (Graber *et al.*, 2009; Endut *et al.*, 2010; Da Silva Cerozi and Fitzsimmons, 2016c). The APIVA® project has provided an opportunity to develop this topic in depth in order to add to our knowledge of the subject and to validate and/or supplement the design data available to date.

Figures 3-1, 3-2 and 3-3 show mass balances carried out on N, P and K nutrients as part of the APIVA® project, on the RATHO experimental pilot in 2017 on a trout/lettuce and lamb's lettuce combination; other balances were also carried out on other macronutrients.

The inputs (represented by the grey arrows on the left) are: fish feed, circulating water present in the system at the start of the experiment and new water added throughout the experiment. The outputs (represented by the white arrows on the right) are: the biomass of lettuce, lamb's lettuce and trout produced, the sludge-sediment-liquidate mixture, the circulating water at the end of the experiment and the water discharged from the system by overflow. The N, P and K inputs are represented by arrows of varying size depending on their importance in the total input: for example, the feed distributed over the test period provided 88.62%, 97.82% and 73.55% respectively of the N, P and K that entered the system, the rest being provided by the new water added throughout the test and the water present in the system at the very start of the experiment: the total represents 100%. The principle is the same for the outputs, and a balance is totally balanced if there is also a sum of 100% at the output of the system. When the balances are not balanced, this is considered to be due to measurement uncertainties for elements such as phosphorus and potassium, which cannot be evacuated in gaseous form, or to degassing for elements such as nitrogen, which can be evacuated in gaseous form.

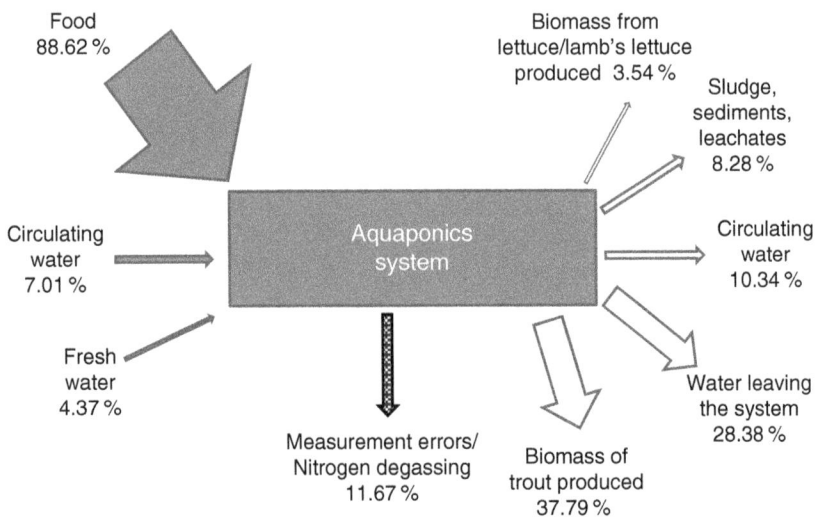

Figure 3-1. Nitrogen mass balance in a trout/salad and lamb's lettuce aquaponic system (ITAVI)

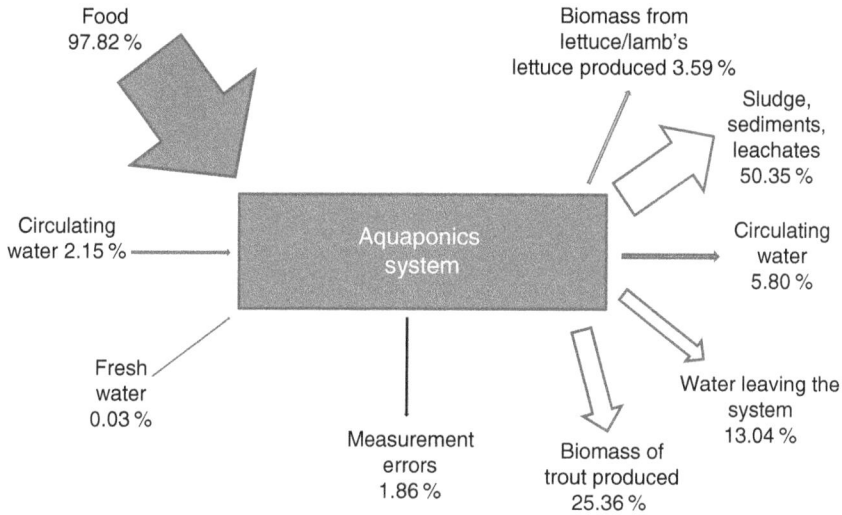

Figure 3-2. Phosphorus mass balance in a trout/salad and lamb's lettuce aquaponic system (ITAVI)

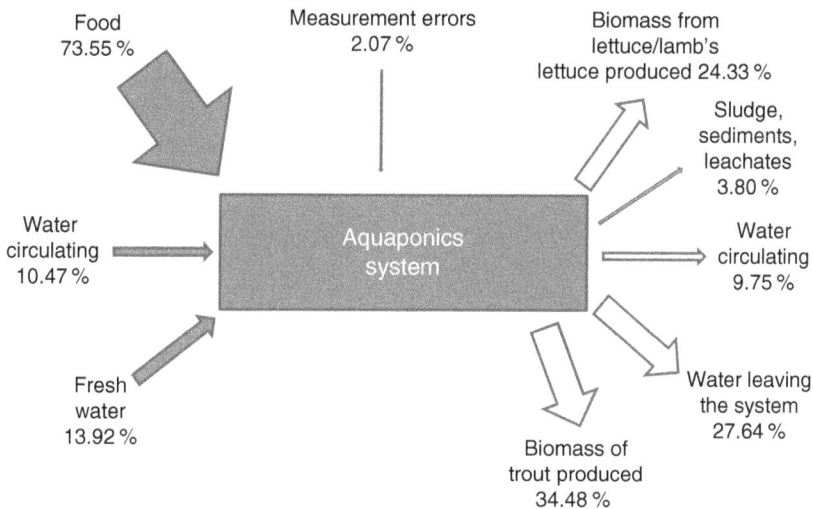

Figure 3-3. Potassium mass balance in a trout/salad and lamb's lettuce aquaponic system (ITAVI)

The results of this mass balance were obtained in a particular context of fish feed/plant surface ratio (90 g of feed at 42% protein/m² of plant surface/day) and fish system opening rate (500 l of new water per kilo of feed distributed), with a particular fish species and plant varieties. It is easy to see from these results that potassium is the most limiting element compared with phosphorus and nitrogen in terms of the plant surface area possible for a given quantity of feed: in fact, the plants consumed 24.33% of the potassium entering the system, while 37.39% of the potassium entering was not consumed (evacuated in the overflow or present in the water at final T). In comparison, only 3.54% of the N and 3.59% of the P entering the system was consumed by the plants, while 38.72% of the N and 18.84% of the P was still available in the crop water and the water discharged from the system. The elements not consumed could theoretically be consumed if there were enough plants. On the basis of this observation, we can estimate that it would have been possible to divide the feed/plant area ratio by 12 for nitrogen, by 5 for phosphorus and by 2.5 for potassium, giving minimum ratios of 13 g feed/m^2/day, 24 g feed/m^2/day and 66 g feed/m^2/day for N, P and K respectively. It is therefore possible to reduce the RASV to around 20 g feed/m^2/day if compensatory potassium is added using pH buffer $KHCO_3$ or potassium sulphate, which makes it possible to install more plant surface for the same amount of feed distributed daily, and thus to better purify the nitrogen and phosphate effluents in the fish compartment. Other as yet unpublished studies have validated the hypothesis derived from this mass balance example, namely that a RASV of 20–25 g/m^2/day was sufficient, provided that additional potassium was added. This ratio could be further reduced by adding phosphorus from the biodigestion of fish sludge from the filtration system, to make this element less limiting.

This type of material balance can be carried out in very different contexts and will give variable results, hence the interest in multiplying them to acquire increasingly reliable dimensioning data that can be adapted to different types of situation.

13.5. Coupled and decoupled systems, what impact on nutrient balance?

In a coupled system, it is easy to imagine that a well-sized plant compartment would provide a certain stability in the level of nutrients in the water from which the fish originate, thus preventing an accumulation of nutrients in the water. However, we have seen the limits of the coupled approach in a large-scale commercial context.

Contrary to popular belief, it is a mistake to assert that a decoupled system leads to an imbalance between the fish and plant compartments, as long as the sizing is adequate. The opening rate of the fish farming system (litres of new water/kg of feed distributed/day) will condition the opening rate of the plant culture system (litres of water from the overflow of the fish farming compartment/day): remember that in a recirculated fish farming circuit, the renewal water volumes are 100 to 400 times lower than in an open circuit, which greatly limits the environmental

impact linked to water use! This opening rate will limit the rate of nitrate and phosphorus accumulation in the fish compartment: a concentration threshold for these elements (in mg/l) will be reached as soon as the quantity (in mg) leaving the fish compartment towards the plant compartment equals the quantity (in mg) entering via the feed. For several weeks after the system is started up, nitrate and phosphorus levels (and other elements) will tend to increase in the rearing water because the quantity of minerals leaving the system will be lower than the quantity entering, until a sufficient concentration is reached in the rearing water to balance the inflow and outflow. The opening rate will condition the maximum concentration of nutrients in the system and allow it to adapt as best as possible to the needs of the plants, while taking care not to reach nitrate toxicity for the fish. If the plant compartment downstream of the fish farming system is sufficiently well sized, the daily dose of minerals leaving the fish farming system can then be consumed, limiting discharges into the environment as much as possible. Any system, whether coupled or uncoupled, generally allows for water renewal and therefore for a certain volume of water to be released through the overflow (which may be the same or much less, depending on the evaporation capacity of the plants, which, let's not forget, consume water). This overflow can be discharged into a watercourse, subject to compliance with environmental standards, or ideally used to irrigate field crops to complete the nitrogen and phosphorus cycle.

14. Sizing aquaponic systems

How can we understand the methods for sizing the various key components of an aquaponics system, and the various hydraulic parameters governing the proper functioning of the biological and physical processes taking place within it?

14.1. Sizing a fluidised bed biological filter

Sizing a biological filter can be done in various ways. The approach described below provides a basis for sizing a fluidised bed biofilter, the most proven and effective technique in recirculating aquaculture.

14.1.1. Estimating the maximum amount of TAN that a fish farming system can generate

TAN is short for *Total Ammonia Nitrogen*. Ammonia nitrogen is a dissolved waste product released by fish during digestion, which is extremely toxic to fish in low concentrations. This is why a recirculating fish farming system must include a biological filter, whose role is to transform ammoniacal nitrogen N-NH4$^+$ (dissolved ionic form) into nitric nitrogen N-NO3$^-$ through the nitrification process that takes place within it. TAN also contains a proportion of ammoniacal nitrogen in gaseous form, N-NH3, which is kept to a minimum due to its high toxicity, and which constitutes a negligible proportion of TAN under the temperature and pH conditions expected in aquaculture and, by extension, aquaponics. The quantity of TAN released by fish is estimated using the following formula (Timmons and Ebeling, 2007): $TAN(g) = \frac{QA \times PB \times 0.092}{t = 1\ jour}$ where QA corresponds to the maximum quantity of feed distributed daily (kg/day); PB corresponds to the proportion of crude protein (% crude protein) contained in the commercial feed; 0.092 is a constant established on the basis that proteins contain 16% nitrogen, 80% of which is assimilated by the organism, and 90% of the nitrogen is excreted in the form of TAN and the remaining 10% in the form of urea. In addition, it takes into account by default that most of the nitrogen contained in faeces and uneaten feed particles is rapidly eliminated from the system by mechanical filtration.

This is the simplest and most accessible formula for determining ammonia emissions. For greater accuracy than the Timmons and Ebeling formula, the complex model developed by INRA can be used in the case of rainbow trout (Papatryphon *et al.*, 2005).

14.1.2. Estimation of the surface area required for the production of nitrifying bacteria

The bacteria in the biofilter responsible for carrying out nitrification need an installation surface area and a physical substrate to develop,

unlike other micro-organisms which can develop freely suspended in the water. The aim is to define a minimum surface area for bacterial installation in line with the TAN discharges from the fish compartment. This surface area (SMI) will vary greatly depending on the physico-chemical parameters of the water. This surface area can be estimated using the following formula:

$SMI = \frac{TAN(g)total}{DNS}$ where DNS corresponds to the degradation potential of TAN per unit area, which ranges from 0.2 to 1 $gTAN/m^2/day$ for water at 15 to 20°C and from 1 to 2 $gTAN/m^2/day$ for water at 25 to 30°C.

14.1.3. Estimate of the minimum reception volume of the fluidised bed biofilter

The volume of media (VM) required to match the SMI calculated above is determined by the specific surface area (SS) of the media used in the biofilter. The specific surface area of bacterial media (or *curlers*) generally ranges from 200 to 1,000 m^2/m^3.

The formula to be used is as follows: $VM = SMI/SS$. A biofilter should contain around 65% media and 35% water to ensure good fluidity of the media through bubbling and a certain level of friction between the *curlers*. This makes it possible to deduce the total volume of the tank making up the biofilter (*VT*) needed to contain the VM: $VT = VM/65\%$.

As theory is not always confirmed by practice, it is recommended and more prudent to overestimate the sizing of the biofilter by at least 20% for greater flexibility and greater adaptability of the biofilter to environmental variations that can have a significant impact on the activity of the bacteria (temperature and pH in particular).

14.2. Sizing the oxygenation system

It's important to bear in mind that a recirculating system involves the rearing of both fish and bacteria, so the oxygenation system must be sized to accommodate both. In addition, even if the plants meet their oxygen requirements via the leaf surface, the roots also need to be immersed in sufficiently oxygenated water to avoid root asphyxia, which is conducive to the development of disease. This compartment therefore needs to be taken into account, but it is simpler to consider the whole system (fish + bacteria) on the one hand, and the plants on the other, independently. It is in fact easy to supply the plants with enough oxygen by bubbling or by means of a sufficiently high circulating flow rate in the culture media.

14.2.1. Estimation of dissolved oxygen requirements for the fish compartment

There are various reference models in the literature that take into account the essential parameters for assessing fish requirements: species, size and water temperature. The Muller-Feuga model, developed for rainbow trout (Belaud, 1996), calculates the dissolved oxygen requirement (MO_2) for one kilo of trout biomass per hour ($mg_{O2}/kg_{fish}/H$) using the following formula: MO_2 (for 1 kg of trout) $= c \times M^a \times 10^b$ where M is the average weight of the fish (in g); the constants a, b and c are governed by the following rules: if $T \, °C \leq 11 \, °C$, $c = 75$, $a = -0.196$ and $b = 0.055 \times T \, °C$; if $T \, °C > 11 \, °C$, $c = 249$, $a = -0.142$ and $b = 0.024 \times T° \, C$.

The value obtained is then multiplied by the maximum biomass of fish (kg) that can theoretically be accommodated by the system to obtain a value for MO_2 (total) in mg O_2/h. When sizing a system taking all these parameters into account, it is preferable to overestimate the water temperature in order to achieve summer conditions, as the higher the temperature, the lower the solubility of oxygen in the water. This method corresponds to the assessment of the needs of rainbow trout: sizing an aquaponics system for this species, which is particularly demanding in terms of water quality - as are most salmonids - ensures a more than adequate supply of oxygen for all other fish species.

A second, simpler and more generalisable method exists (Timmons and Ebeling, 2007). It takes into account only the fish biomass (kg) and the daily feeding rate: MO_2 *(for 1 kg of fish)* $= SP \times TA \times aDO$ where *SP* is the maximum fish biomass in the system (kg); *TA* is the daily feeding rate (kg feed/kg fish/day); *aDO* corresponds to a ratio between the quantity of oxygen to be supplied and the quantity of feed distributed and varies between 0.25 and 0.5 kg O_2/kg feed for a herd of aquaculture fish in the grow-out phase (Belaud, 1996; Timmons and Ebeling, 2007). This value should be adapted according to the species (rather the high range for very demanding fish such as salmonids, and rather the low range for less demanding fish such as tilapias or cyprinids) and the growth stage of the fish.

14.2.2. Estimation of dissolved oxygen requirements for the bacterial compartment

According to the literature, nitrifying bacteria need around 4.57 g of oxygen to break down 1 g of ammoniacal nitrogen (Pambrun, 2005). According to Timmons and Ebeling (2007), a ratio of 0.12 kg of oxygen per kg of feed distributed is sufficient for nitrifying bacteria.

It is also necessary take into account the activity of the heterotrophic bacteria present in the system, even if we try to avoid this as much as possible with efficient mechanical filtration. These bacteria are thought to consume between 0.13 and 0.5 kg of oxygen per kilo of feed distributed in the system, which is far from negligible. It is reasonable to multiply the requirements of nitrifying bacteria by three to take account of heterotrophic activity (Timmons and Ebeling, 2007).

14.2.3. Recirculation flow calculation

The water must be able to flow through the fish farming system at a rate that satisfies the oxygen requirements of the fish and bacteria, but also allows all solid (through mechanical filtration), dissolved and gaseous (through biological filtration and water aeration) waste to be removed.

The circulating flow rate (CFR) to be adopted to circulate the water in the recirculated aquaculture system can be evaluated as a function of the estimated dissolved oxygen requirements in the system (for the fish and for the nitrifying bacteria) and the minimum concentration that we wish to maintain (7 mg/l). To do this, it is necessary to take into account the species raised and the maximum biomass in stock, and the maximum oxygen concentration in the water, which is a function of temperature and altitude.

The circulating flow rate can be determined by taking into account the notion of hourly renewal in the basins (renewal rate or RR). Two to four water changes per hour in each tank is ideal to ensure effective mechanical filtration and good aeration (Timmons and Ebeling, 2007; Trang *et al.*, 2017). It is therefore sufficient to multiply the rearing volume (m³) by the hourly renewal rate (number/H) to estimate the sufficient DC (m³/H): *DC = Rearing volume× hourly renewal rate.*

14.3. Dimensioning hydraulic parameters

14.3.1. Hydraulic retention time in the biofilter

By calculating the biofilter reception volume VT and the circulating flow rate DC, we can then calculate the water residence time TS (or retention time) in the biological filter, which determines the time available for the bacteria to break down the dissolved elements. We aim for a TS of at least 5 minutes: *TS = VT/DC.*

The TS is then expressed in hours; simply multiply it by 60 to obtain the result expressed in minutes. If the TS is less than 5 minutes, you will have to decide whether to oversize the biofilter volume and/or reduce the recirculation flow rate until an adequate value is reached.

14.3.2. System opening rate

A recirculated system is rarely completely closed, to avoid excessive accumulation of fine particles, nitrates and other minerals, and also to limit the risk of *off-flavour* in fish flesh. This is why the water is partially renewed and a system opening rate is set.

An order of magnitude can be estimated for the renewal of new water depending on the quantity of feed distributed: carp and tilapia can be satisfied with a renewal of water varying between 0.1 and 0.3 m³ of new water/kg of feed/day, whereas trout will have requirements varying between 0.2 and 1 m³ of water/kg of feed/day depending on the technical nature of the systems.

Figure 3-4 summarises the different intensities of recirculation in fish farming systems, with the different arrangements required to allow the system to be closed at different rates. The more closed the circuit, the lower the amount of new water required to treat the farm effluent, and the more it will be necessary to compensate for this low level of water renewal with additional equipment (oxygenation, degassing, mechanical and biological filtration, UV, infrastructure, etc.).

14.3.3. Hydraulic retention time (HRT) or residence time in the hydroponic compartment

The residence time of the aquaponic nutrient solution in the plant compartment must be sufficient for the macro- and micronutrients present in the livestock effluent to be effectively absorbed by the plants. The hydraulic retention time for a system on an inert substrate is calculated as follows: $TRH = \frac{As \times hw \times \phi}{DC}$ (Endut *et al.*, 2010; Connoly and Trebic, 2010) where *As* corresponds to the surface area of the crop (m²); *hw* corresponds to the height of water in the crop structures (m); ϕ corresponds to the porosity of the media (unitless index varying between 0 and 1): for a *raft* or NFT type system, ϕ is taken to be *1* as there is no media; *DC* corresponds to the circulating flow rate (m³/h).

If the HRT is too high, there is a risk of root asphyxia due to insufficient water renewal in the growing zones. An HRT that is too low may reduce the efficiency of phytodepuration through root extraction (Connoly and Trebic, 2010; Endut *et al.*, 2010).

According to Endut *et al* (2009), the ideal HRT would be around 35 minutes for a system on an inert substrate (MFG), with an RASV of 15 to 42 g of feed/m² of plant surface/day, with a catfish/water spinach coupling. According to the characteristics of the UVI system, the target HRT is 3 to 4 hours for a system with an RASV of 60 to 100 g feed/m² plant area/day with a tilapia/bratfish combination (Rakocy *et al.*, 2006). It is difficult to compare the few publications available on the subject, given that :

– feed/area ratios are not always specified;
– the concentrations of nutrients in the water are not the same;
– the plant species tested are not the same.

It should be noted that *raft* systems can afford to have a higher HRT because of the large volume of water stored in the culture basins and the low risk of root asphyxia. According to these results, *raft* systems require a longer retention time than inert substrate systems. During experiments carried out as part of the APIVA® project, flow rates of between 0.25 and 1 renewal per hour in the *rafts* proved sufficient (water retention time of 1 to 4 hours) for plant growth and oxygenation of the environment. The Inra-Peima experimentation

RECIRCULATION INTENSITY OF FISH FARMING SYSTEMS

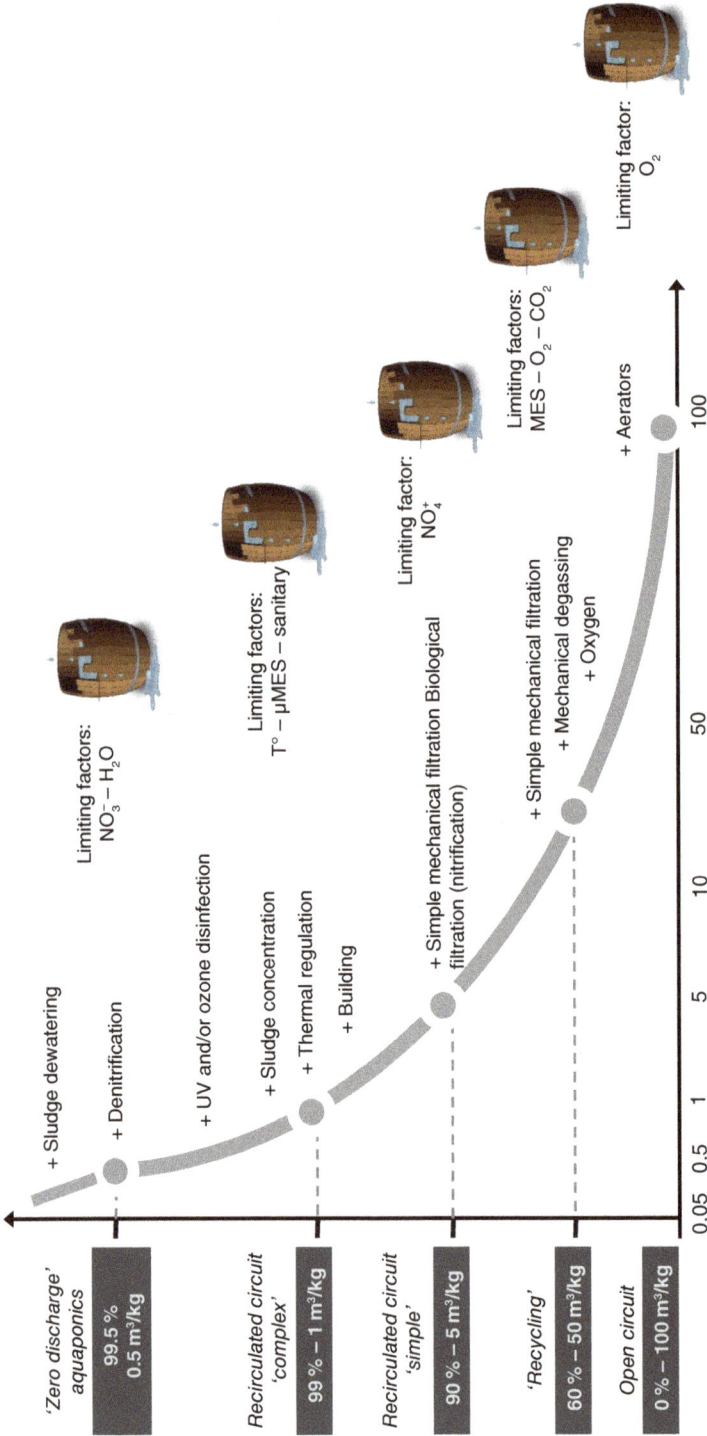

Figure 3-4. Different recirculation intensities in fish farming systems. (Matthieu Gaumé, ITAVI)

station has even grown plants on *rafts* with 0.5 water changes per day, with no problems associated with asphyxiation of the growing medium.

14.3.4. Volume of water to be renewed in the hydroponic compartment

The evapotranspiration (ET) of water by plants constitutes the main consumption of water in the plant compartment. The water demand of plants under shelter depends largely on the solar radiation captured by the crop, and therefore on the energy received and the leaf surface area. The Villèle formula developed by INRA estimates the potential evapotranspiration in a greenhouse (ETPs, expressed in litres of water/m² of crop or in mm) (CTIFL, 1995) and is applicable when the plant cover is continuous. This formula is expressed as follows $ETPs(l/m^2) = 0.67 \times \frac{RG \times C}{l}$ where the constant 0.67 comes from the fact that approximately 67% of the solar energy reaching the plant is used for transpiration; *RG* corresponds to the global external radiation, which is estimated at 2.5 MJ/m²/day for a sunny day; *C* corresponds to the transmission coefficient depending on the material making up the greenhouse (glass, single or double-walled plastic): 0.7 for a single-walled plastic, 0.65 for a double-walled plastic, 0.75 to 0.8 for a glass wall; *l* corresponds to the latent heat of vaporisation of water, which is 2.51 MJ/l of water at 15–20°C. Depending on the stage of the crop, a cultural coefficient K is applied to the ETPs to determine the real evapotranspiration (ETR), taking into account the growth stage of the plant: *AET = ETPs× K* where : *K* is a coefficient that varies between 0.1 (for plants at juvenile stages) and 1 (for a plant at harvest stage). So, water requirements will depend directly on the amount of sunshine, and therefore on the season, but also on the growing stages of the plants.

There are many formulas for approaching the measurement of evapotranspiration, but they are often complex. Villèle's formula has the merit of being simple, but does not take into account many important variables. The most accurate - and most complex - formula is that of Penman and Monheith, which incorporates numerous parameters that can be measured or calculated from meteorological data (variations in temperature, humidity and atmospheric pressure, latitude, altitude, sunshine duration and wind strength) and agronomic data (albedo, plant stomatal conductivity, plant height, type of substrate or soil, etc.).

In practice, the daily renewal in the fish compartment more than compensates for water losses through evapotranspiration and the overall result is a virtually closed circuit, with volumes of wastewater lower than volumes of 'new' water due to losses through evapotranspiration and evaporation. It is therefore not compulsory to determine water losses by these means when sizing an aquaponics system.

14.4. Sizing a recirculation pump and hydraulic circuit

Pumps transfer energy between a mechanical device and a fluid. They communicate potential energy to the fluid by increasing the pressure downstream, and kinetic energy by setting the fluid in motion. There are many types of pump: "positive displacement" pumps (diaphragm, lobe, piston, vane, gear, screw, peristaltic, grain, rotary barrel, rotary plate, etc.) and "turbopumps" (centrifugal radial pumps, helical centrifugal pumps, axial propeller pumps, multi-stage turbine pumps, etc.).

The flow rate and pressure characteristics of a pump determine an operating curve that can be compared with the head losses and gradient of the network. This makes it possible to determine an operating point for the pump/network system, i.e. to find a flow rate (Q) that balances the pressure supplied by the pump or, conversely, the pressure required to operate the network at this flow rate. The flow rate/pressure ratio defines a hydraulic power P, calculated as follows: *P = Q × density of the fluid × total head (TH)*, where the density of water is approximately 1,000 kg/m³; the flow rate in m³/h that we want to ensure in the system; the total head (TH) expressed in metres of water column mCE.

The total manometric head (HMT) for a fish-farming installation is calculated by adding together the difference in level of the pumping installation and the head losses:

– The difference in altitude of the pumping installation is a major pressure consumer. It is the difference in altitude between the liquid inlet and its outlet to the atmosphere. The gradient is made up of the "geometric suction head" and the "geometric discharge head"; the first representing the difference in level between the lowest water level and the pump axis, and the second being the difference in level between the pump axis and the highest point of the distribution, all expressed in metres;

- Pressure losses occur when water is transferred through pipes. As the water circulates through the pipes, it rubs against the walls, consuming part of the pressure supplied by the pump; these are the famous head losses. The smaller the cross-section of the pipe, the greater the pressure losses. This loss of energy, linked to the speed of the fluid (low speed = low pressure drop), is caused by the transformation into heat of internal friction caused by the viscosity of the fluid (a perfect fluid with no viscosity generates no pressure drop), the roughness of the walls, variations in speed and variations in the direction of the fluid.

There are two types of pressure drop:

- Regular head losses, linked to friction in pipes of constant cross-section. These are caused by the viscosity of the fluid. The flow rate and the resulting pressure losses are used to choose the right diameter for a pipe, depending on the material. The pressure drop is proportional to the length of the pipe, which is why this type of pressure drop is expressed in metres per 100 metres of pipe length (m/100 m). Linear head losses of around 3 m/100 m are generally accepted as a good economic compromise;

- singular head losses, which are the result of variations in fluid velocity and changes in direction caused by shapes and obstacles encountered by the fluid as it passes through an object: bends, connections, junctions, valves, etc. They are expressed in pascals or metres of water column (mCe). They are expressed in pascals or metres of water column (mWC). Singular head loss coefficients (λ) are given for these "obstacles". These data are generally supplied by the manufacturers of hydraulic accessories. Certain formulas can be used to calculate these coefficients, such as the Weisbach formula for bends or the Lorenz formula for diverging cones.

In reality, these two types of head loss (regular and singular) are not always separate. For example, in a rounded bend, there is a proportion of singular head loss due to the change in direction and a proportion of regular head loss due to friction along the length of pipe formed by the bend. The two head losses may need to be added together if the friction surfaces are large, but in general the regular losses are neglected compared with the singular losses. As the pressure losses in the hydraulic circuit are approximately proportional to the pump flow rate squared (Q^2), the shape of the $HMT = f(Q)$ curve for the hydraulic circuit is parabolic.

Once the required flow rate and total head have been estimated, the characteristic curves $HMT = f(Q)$ of the various pump models in the same series can be seen in the charts supplied by the suppliers. To determine the most suitable model, you need to place the HMT/flow rate pair on the chart to define its 'operating point'. This allows you to choose the pump whose characteristic curve is immediately above this point. A safety margin of 10 to 20% is generally allowed when sizing the pump.

15. The physico-chemical parameters of water: a history of compromise

As we have seen, it is vital to correctly size the system components (rearing tanks, biological and mechanical filtration, oxygenation, plant surface, etc.) and the hydraulic parameters (renewal rate, opening rate, retention time, etc.) governing the circulation of water in these components.

Once a functional system is in place, another problem emerges: water management. Water is central to the operation of an aquaponics system: it is necessary for the life of the fish, bacteria and plants. Over time, water changes from a physico-chemical point of view, and it is vital to understand the mechanisms and issues involved in these changes in order to interpret and manage them on a day-to-day basis.

The combination of hydroponics and aquaculture requires water quality parameters to be reconciled for the joint survival and growth of plants, fish and nitrifying bacteria. To achieve this, it is essential to control a large number of parameters, the most critical being the concentration of dissolved nitrogen compounds, pH, temperature, dissolved oxygen concentration, organic matter load and conductivity. These parameters must be rigorously controlled, while sound sizing must be carried out before a system is designed to ensure irreproachable mechanical and biological filtration, as well as sufficient circulation and oxygenation of the water.

15.1. Concentration of dissolved nitrogen compounds

The concentrations of nitrogen molecules (ammonium, nitrites, nitrates) are expressed in different ways in the literature. It is important to understand the meaning of these different notations, as their misinterpretation is often a source of error in assessing the actual mineral nitrogen load in water. Nitrate concentration is generally expressed as $N-NO_3^-$ (nitric nitrogen), which corresponds to the nitrogen contained in the NO_3^- nitrate molecule, without taking into account the oxygen atoms that make up the nitrate molecule. This expression is therefore called "in N". Similarly, $N-NH_4^+$ corresponds to the nitrogen contained in the ammonium ion (ammoniacal nitrogen), and $N-NO_2^-$ corresponds to the nitrogen contained in the nitrite ion (nitrous nitrogen). So, for example, one gram of NO_3^- ion corresponds to 0.226 grams of nitric nitrogen and conversely one gram of nitric nitrogen corresponds to 4.427 grams of NO_3^- ion. Table 3-1 details the conversion constants used to convert from the "ionic" expression to the "N" expression.

15.1.1. Ammoniacal nitrogen: $N-NH_3$ and $N-NH_4^+$

The intensification required to achieve economic profitability in recirculated aquaculture

Table 3-1. Conversion constants for nitrogen parameters (ITAVI)

Concentration conversion of nitrogen molecules (mg/l)

Change from "ionic" to "N" expression	Switching from "N" to "ionic" expression
$[NO_3^-] \times 0.226 = [N-NO_3^-]$.	$[N-NO_3^-] \times 4.427 = [NO_3^-]$.
$[NO_2^-] \times 0.304 = [N-NO_2^-]$.	$[N-NO_2^-] \times 3.284 = [NO_2^-]$.
$[NH_4^+] \times 0.776 = [N-NH_{(4)}]^+$	$[N-NH_4^+] \times 1.288 = [NH_{(4)}]^+$

systems generates substantial ammoniacal nitrogen production. Ammoniacal nitrogen exists in two forms in water: ammonia (NH_3, a dissolved gas in non-ionised form) and ammonium ion (NH_4^+, a mineral element in ionised form).

The non-ionised form (also known as "undissociated") is the most toxic for fish. It causes irritation and solidification of the gill lamellae, leading to a reduction in the absorption surface of the gills, which can cause respiratory problems in fish. The ammonia environment is also conducive to the development of fish pathogenic myxobacteria. The NH_4 form[+] is less toxic but still dangerous, and tolerance varies greatly from species to species.

It is also often referred to as Total *Ammonia* Nitrogen (TAN), which is the sum of $N-NH_3$ and $N-NH_4^+$ (Kinkelin *et al.*, 1985; Losordo *et al.*, 1998). The proportion of $N-NH_3$ in TAN increases as the pH and temperature of the water rise. For the same temperature, a very small variation in pH can have a very large impact on the proportion of $N-NH_3$ in TAN, as shown in Figure 3-5.

Generally speaking, for salmonids, the NH_3 concentration should be kept below 0.025 mg/l NH_3 (i.e. 0.020 mg/l $N-NH_3$), as mortality is likely to occur above these thresholds in the short term (Losordo *et al.*, 1998; Timmons and Ebeling, 2007) and it is even advisable to aim for a maximum threshold of 0.01 mg/ $N-NH_3$. Warm-water fish are more tolerant. Tilapia in particular can tolerate concentrations of 0.1 to 0.2 mg/l of $N-NH_3$ (Treadwell, 2010).

Tests to measure ammoniacal nitrogen levels in water usually give results expressed as $N-NH_4^+$ or NH_4^+. A reasonable target for aquaculture is to remain below 1 mg/l NH_4^+ (0.77 mg/l $N-NH_4^+$) for cold water fish (and preferably below 0.5 mg/l $N-NH_4^+$ over time) and below 2.5 mg/l NH_4^+ (2 mg/l $N-NH_4^+$) for warm water fish.

Table 3-2 gives the maximum concentrations (mg/l) of total ammoniacal nitrogen (TAN) according to water temperature and pH, so as not to exceed the threshold of 0.01 mg/l of $N-NH_3$ for cold-water fish such as salmonids (between 0 and 20°C).

15.1.2. Nitrites

Nitrite (NO_2^-) is the intermediate form in the nitrification of ammoniacal nitrogen into nitrate.

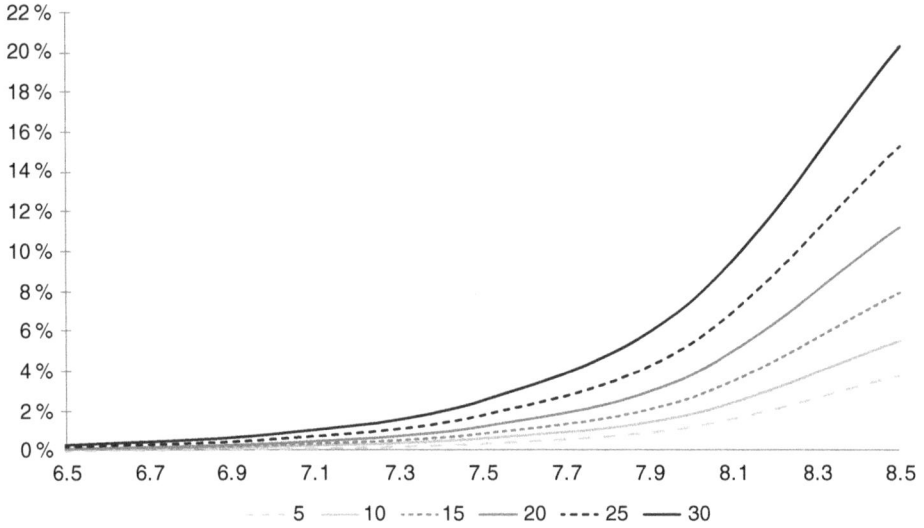

Figure 3-5. Share (in %) of N-NH$_3$ in TAN (N-NH$_3$ + N-NH$_4^+$) as a function of temperature and pH. (ITAVI, adapted from Emerson *et al.*, 1975)

Table 3-2. Maximum concentrations (mg/l) of TAN acceptable according to water temperature and pH to avoid toxicity phenomena linked to the N-NH$_3$ form, for salmonids. (ITAVI)

Temperature (°C)	pH				
	6.5	7	7.5	8	8.5
5	25.30	8.00	2.50	0.81	0.26
10	17.00	5.40	1.70	0.55	0.18
15	11.60	3.70	1.20	0.38	0.13
20	8.00	2.50	0.80	0.26	0.09

Nitrites affect the function of fish gills, disrupting their ability to transfer oxygen into the bloodstream by altering the conformation of the haemoglobin molecule (Dolomatov *et al.*, 2013). A temporary malfunction of the system, or a drop in the activity of nitritating bacteria, can lead to NO$_2^-$ peaks. Fish tolerance to nitrite toxicity is highly variable, depending on the duration of exposure, water quality, species and size of individuals, and the level of chlorine in the water, which has an oxidising effect on nitrite (Kroupova, 2005). Cyprinids (carp) and tilapia are much more resistant than salmonids (salmon, trout, etc.). In addition, juveniles are generally more tolerant than adults. The lower the oxygen saturation level, the greater their toxicity

(Pillay and Kutty, 2005). This type of problem can only be avoided by professional design of the biological filter. You should always aim for a concentration of less than 0.2 mg/l of N-NO$_2^-$ in cold water, and less than 0.5 mg/l of N-NO$_2^-$ in warm water.

15.1.3. Nitrates

A concentration of around 100 to 250 mg/l of N-NO$_3^-$ is recommended (Resh, 2004 in Tyson, 2008) for hydroponics. In practice, a concentration of between 30 and 60 mg/l of N-NO$_3^-$ is quite sufficient for the majority of plants grown in aquaponics.

A recent study (Davidson, 2014) shows that a nitrate concentration of 80 to 100 mg/l N-NO$_3^-$ would be at least partially responsible for chronic effects (reduced survival, changes in swimming behaviour) on the health of juvenile rainbow trout. Therefore, as a precautionary measure and in the absence of detailed data, we can aim for a threshold of 180 mg/l of N-NO$_3^-$ or even more for resistant species such as tilapias and cyprinids, and 100 mg/l of N-NO$_3^-$ for sensitive fish such as salmonids.

For plants, the NO$_3^-$/NH$_4$ ratio$^+$ within the hydroponic compartment should ideally vary between 100%/0% and 90/10%, depending on the species and the growth phase (flowering, fruiting): a higher proportion of ammonium ion can lead to root toxicity, while the absorption capacity of certain minerals - such as K^+, Ca^{2+}, Mg^{2+} - risks being inhibited (Cockx and Simonne, 2003; Tyson *et al.*, 2004). In aquaponics, the presence of ammoniacal nitrogen in the water of the plant compartment is unlikely if a biofilter is correctly sized, which is a *sine qua non* condition for successful fish farming. The NO_3^-/NH_4 ratio[+] is therefore closer to 100%/0%.

One of the challenges of aquaponics is to design a plant compartment capable of reducing the nitrate load in fish farming effluents, so as to limit excessive accumulation and the discharge of nitrogenous effluents into the environment.

15.2. Physico-chemical parameters of the water

15.2.1. The pH

pH is a measure of the acidity of water.

15.2.1.1. BIOFILTER COMPARTMENT. pH is a major factor in regulating the nitrifying activity of bacteria. The optimum pH for nitrifying bacteria is between 7.5 and 9, with a general consensus value of 8.2 (Pambrun, 2005; Rakocy *et al.*, 2006). Figure 3-6 illustrates that the nitrification rate is maximised in this pH range.

Nitrification in the biological filter itself causes a drop in pH in the environment, as this reaction consumes alkalinity in the form of carbonate ions CO_3^-. If the pH is too low, the bacterial activity responsible for nitrification will slow down or even stop, and ammonia compounds will rapidly accumulate in the farm water to toxic levels. To counteract this effect, it is necessary to adjust the alkalinity of the system by adding weak basic compounds such as calcium carbonate ($CaCO_3$) and/or potassium bicarbonate ($KHCO_3$), which have a buffering capacity and stabilise the pH at a value close to neutrality, while enriching the medium in potassium or calcium, present in the nutrient solution in quantities that are sometimes limiting for plant growth. Strong bases such as calcium hydroxide ($Ca[OH]_2$) and/or potassium hydroxide (KOH) can also be added to raise the pH quickly (Rakocy *et al.*, 2006; Lennard and Leonard, 2006), but their use is more risky due to the risk of rapid and uncontrolled changes in pH in

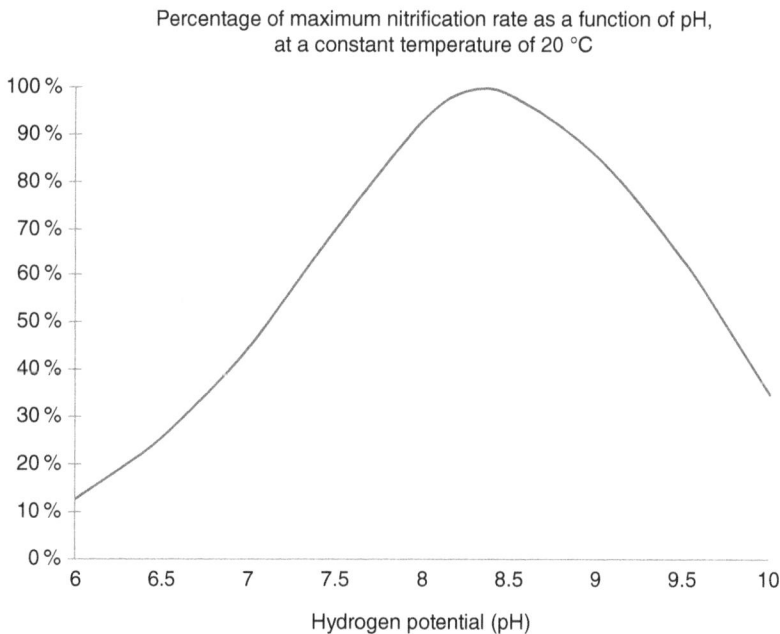

Percentage of maximum nitrification rate as a function of pH, at a constant temperature of 20 °C

Hydrogen potential (pH)

Figure 3-6. Changes in maximum nitrification rate (% of maximum nitrification) as a function of pH at constant temperature (20°C) (ITAVI)

the water. Sodium bicarbonate ($NaHCO_3$) is a source of alkalinity widely used in recirculated aquaculture, but should not be used as such in aquaponics: there is a risk that sodium will accumulate to phytotoxic levels (over 30 mg/l Na^+) and interfere with the uptake of K^+ and Ca^+ (Rakocy et al., 2006), resulting in reduced plant growth. The exact amount of bicarbonate to add to regulate the pH can be determined empirically, depending on the pH and alkalinity of the water used to renew the system and the opening rate chosen (l of new water/kg of feed distributed). In general, if the water source is acidic to neutral (6.5 to 7), bicarbonate should be added at a rate of 15 to 25% of the mass of feed distributed per day, preferably in a recovery tank before being returned to the biofilter or in the biofilter itself, to give the product time to dissolve. Some authors estimate the quantity of buffer to be added directly as a function of the quantity of feed distributed: from 150 g to 190 g of bicarbonate per kg of feed consumed by the fish (Davidson et al., 2011) depending on the protein content of the feed: the higher the protein content, the greater the nitrification required to treat discharges with a fixed quantity of feed, and the greater the compensation required by alkaline elements. In practice, when slightly basic renewal water (7.5 to 8) is used, there is no need to add pH buffers, as the new water provides sufficient alkaline compounds to feed the bacteria. On the other hand, if neutral to acid renewal water is used, the consumption of bicarbonate will in itself constitute a non-negligible production cost. It is also possible to use limestone gravel filters to raise the pH of the water without adding alkaline compounds. Inra-Peima has successfully set up this type of system downstream of the plant compartment, enabling:

–   to recirculate more alkaline water, thereby greatly limiting external inputs of alkalinising compounds;
–   mechanical filtration of fine organic particles passing through plant crops.

15.2.1.2. AQUACULTURE COMPARTMENT. Fluctuations in pH are a major stress factor and should be avoided as far as possible. A pH in the range 6.5 to 8.5 is ideal for most farmed fish. A pH of around 7 is preferable, as the proportion of NH_3 gas in total ammoniacal nitrogen increases proportionately with the rise in pH. The most dangerous thing for fish (apart from extremely acidic or extremely basic pH levels) are sudden changes in pH, which can be fatal for the least tolerant fish, particularly salmonids.

15.2.1.3. PLANT COMPARTMENT. Most of the time, plants have optimum growth potential when the irrigation water has a pH of between 5.5 and 6.5 (Jones, 2005; Kane, 2006; Hochmuth, 2012). The closer the pH of the nutrient solution is to this range, the better the chances of success for soilless plant cultivation. The pH of the water affects the solubility of nutrients (particularly iron, manganese, zinc, copper and boron), making them less available to plants when the pH exceeds 7, while the solubility of phosphorus, calcium, magnesium and molybdenum tends to decrease when the pH falls below 6 (Treadwell et al., 2010; Hochmuth, 2012).

Figure 3-7 shows the evolution of the bioavailability of different nutrients to plants in soil and in hydroponic nutrient media, as a function of pH, according to work carried out by Truog in 1947. Each nutrient is represented by a band whose width is proportional to its bioavailability (Jones, 2005; Trejo-Tellez et al., 2012). A pH that is too low (< 5.5) will cause a drop in the availability of calcium (Ca), magnesium (Mg) and molybdenum (Mb). A pH that is too high (> 7.5) will reduce the availability of iron (Fe), potassium (K), manganese (Mn), zinc (Zn), copper (Cu) and boron (Bo).

15.2.1.4. COMPROMISE IN AQUAPONICS. Given that the pH tolerance range for fish is generally between 6.5 and 8.5 depending on the species, that hydroponically grown plants prefer a pH between 5.5 and 6.5 and that nitrifying bacteria function optimally at a pH of 7.5 to 9, the compromise that seems most suitable is to aim for a pH between 6.5 and 7.5 to promote root uptake of nutrients by plants while limiting the drop in biological filtration efficiency (Tyson et al., 2004; Rakocy et al., 2006; Da Silva Cerozi and Fitzsimmons, 2016b). The lower limit of this threshold favours nutrient uptake by plants, while the upper limit favours bacterial nitrification to the detriment of nutrient uptake by plants. It is often useful to oversize a biological filter on the basis that pH conditions in aquaponics are unfavourable to plant growth. Here we can see the potential benefits of a decoupled system in terms

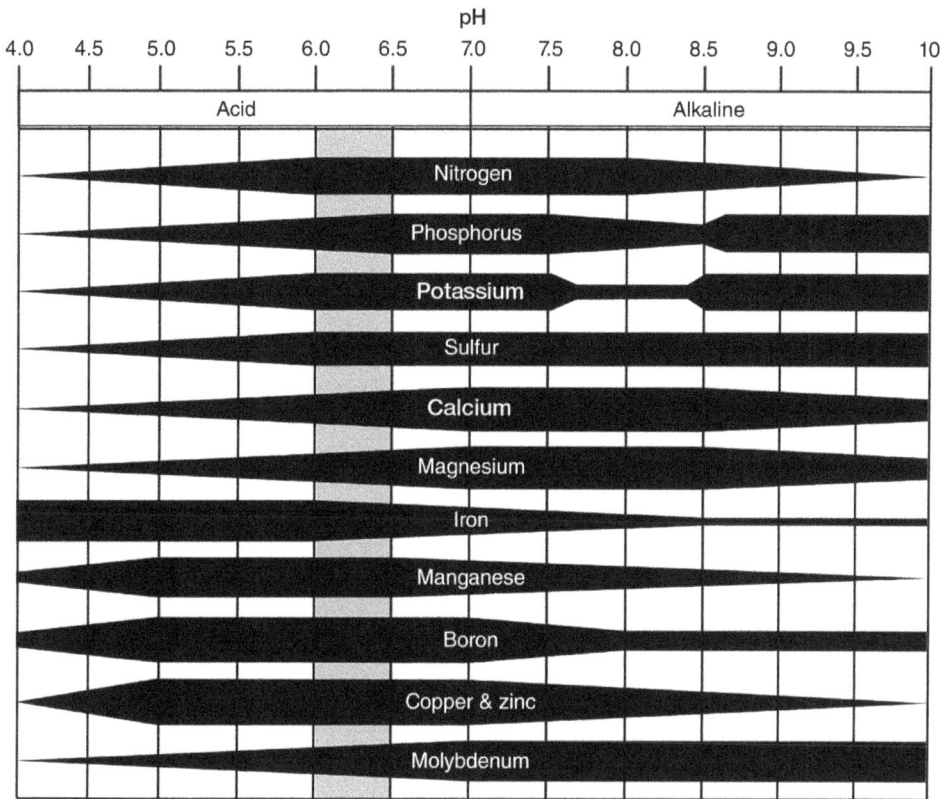

Figure 3-7. Truog diagram showing relative estimates of the biodiponisility of different macro- and microelements for roots, depending on the pH level of the medium (CoolKoon - Own work, CC BY 4.0, based on the work of Truog, 1947).

of controlling the physico-chemical parameters of the water: adding acidic compounds can help manage the pH in the plant compartment without affecting the fish compartment.

These pH recommendations for the plant compartment are based on experience gained in hydroponics. In practice, many plants grown in aquaponics grow very well in water with a pH of around 7.5-8, which calls into question knowledge taken for granted in conventional soilless culture. To what extent do the bacterial populations present in aquaponic solutions have a beneficial effect on nutrient uptake by roots, even in alkaline conditions? This type of research requires the skills of microbiologists and metagenomics specialists.

15.2.2. The temperature

15.2.2.1. BIOFILTER COMPARTMENT. Nitrifying bacteria (e.g. *Nitrobacter* and *Nitrosonomas*) have an optimal growth temperature between 20 and 30°C (Pambrun, 2005). Their growth rate can fall by 50% between 15 and 20°C, and by up to 75% between 8°C and 15°C; their activity is totally inhibited at 4°C, and they die at 0°C or above 49°C. The rate of nitrification is highly dependent on pH - as we saw earlier - but temperature is also a very important factor of variation, as shown in Figure 3-8.

15.2.2.2. AQUACULTURE COMPARTMENT. Each species of fish is characterised by a temperature range within which individuals can remain healthy, and a narrower, optimal temperature range within which they can develop and grow to their full potential. These ranges vary from species to species (table 3-3). It is important to avoid sudden changes in water temperature, and to make sure beforehand that a species will be able to survive all year round in the production area, depending on seasonal temperature variations.

Percentage of maximum nitrification rate
depending on temperature and at optimal pH

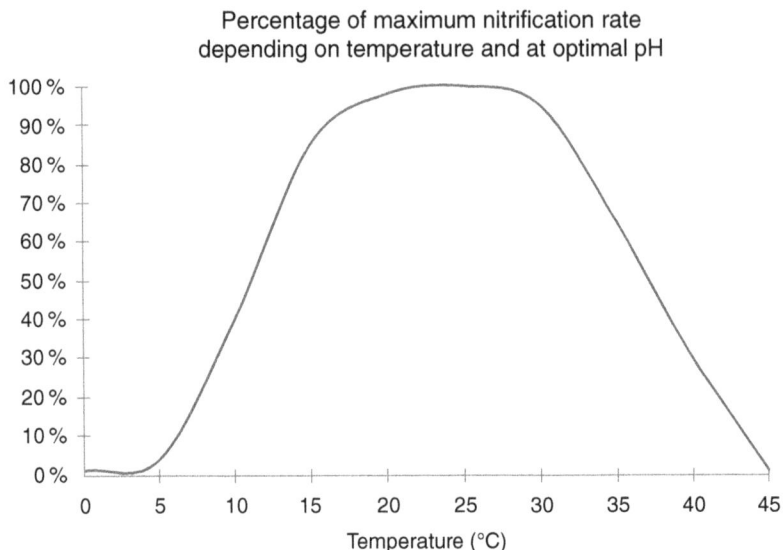

Figure 3-8. Changes in the maximum rate of nitrification (% of maximum nitrification) as a function of temperature and optimum pH. (ITAVI)

Table 3-3. Temperature tolerance ranges for different fish species. (Data from FAO 2014, Cultured Aquatic Species Fact Sheets)

Species	Lower temperature limit (°C)	Optimum growth temperature (°C)	Upper temperature limit (°C)
Brook trout	14	20	24
Brown trout	7	15	21
Rainbow trout	5	14	21
Black bass	10	21	29
Common carp	3	23	35
Common perch	3	22	/
Catfish	13	28	38
Tilapia	17	28	38

15.2.2.3. PLANT COMPARTMENT. As with fish, each plant species has an optimum temperature value and a tolerance range (Trejo-Tellez *et al.*, 2012). Temperature has a direct impact on germination and plant growth. The ideal temperature for many plants, particularly flowers and warm-season vegetables and herbs, is between 15 and 24°C (tomatoes, basil, etc.), while more tolerant vegetables can make do with water at between 10 and 20°C (lettuces, spinach, lamb's lettuce, cabbage, broccoli, etc.) (Rakocy *et al.*, 2006).

15.2.2.4. COMPROMISE IN AQUAPONICS. To meet the requirements of all the organisms present in aquaponics, the water temperature should always be above 12°C, up to a limit of 25–30°C for warm-climate plants and fish. The choice of plants, in particular, should be made judiciously in relation to the temperature that can be maintained in the growing environment, depending on whether the system is under cover in a closed greenhouse or outdoors.

15.2.3. Dissolved oxygen concentration

The rate of saturation of the water with dissolved oxygen decreases over time in aquaculture with respiration by fish and bacterial nitrification reactions, and increases with the addition of oxygen through aeration or oxygenation of the water. When the water is saturated with oxygen, and at a given pressure, temperature has an impact on the maximum dissolved oxygen content that the water can contain: the higher the temperature, the lower the solubility and therefore the availability of oxygen in the water (figure 3-9).

15.2.3.1. BIOFILTER COMPARTMENT. Dissolved oxygen is used as a final electron acceptor by the nitrifying bacteria to carry out the nitrification reactions. The nitrification rate is maintained at 100% for a dissolved oxygen concentration greater than or equal to 3 mg/l (Pambrun, 2005), and more generally for a water oxygen saturation rate of around 80% (Rakocy et al., 2006), provided the temperature and pH are adequate. Below 2 mg/l, another type of bacteria can develop under anaerobic conditions, responsible for the denitrification process, which converts nitrates into atmospheric nitrogen that cannot be used by plants (Somerville et al., 2014). It may be worthwhile for large-scale systems with little water exchange to install an independent denitrifying basin (closed, in anoxic conditions, with the possibility of coupling/decoupling with the system) to avoid too much nitrate accumulation, in case the plants alone are not sufficient.

15.2.3.2. AQUACULTURE COMPARTMENT. For optimum growth, the ideal concentration of dissolved oxygen is 5 to 7 mg/l for warm-water species and 7 to 11 mg/l for cold-water species. The level of oxygen supersaturation in the water should not exceed 120–140%, as this can lead to oxidative stress that can cause death.

15.2.3.3. PLANT COMPARTMENT. Plants use oxygen to carry out cellular respiration, a series of reactions during which plants break down carbohydrates to create energy (ATP). If root aeration is insufficient, a relative increase in CO_2 concentration occurs. Oxygen levels below 4 mg/l can inhibit nutrient uptake by roots, leading to asphyxiation and plant dieback (Gislerød and Kempton, 1983). Furthermore, in an oxygen-poor environment, opportunistic diseases caused by pathogenic fungi can infest the roots. It is therefore necessary to stir the culture water and/or ensure a sufficient renewal flow rate, in the case of *raft* culture: a water residence time of 1 h to 4 h in *raft* culture shelves is largely satisfactory, while a flow rate circulating under the roots of 0.8 to 8 l/min seems to give good results (Maucieri et al., 2017).

15.2.3.4. COMPROMISE IN AQUAPONICS. An oxygen level of 7 mg/l is suitable for the most demanding

Change in oxygen saturation level as a function of water temperature

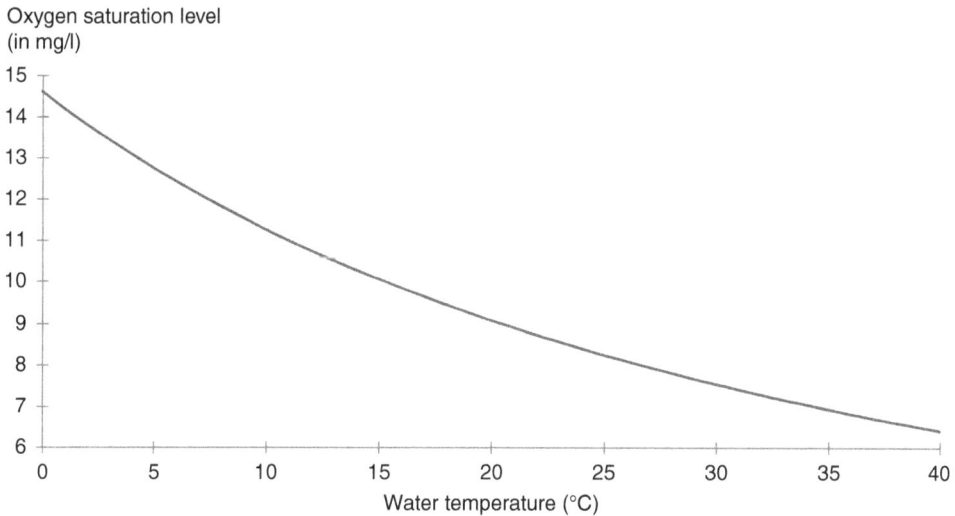

Figure 3-9. Changes in oxygen levels in water as a function of temperature, at maximum water oxygen saturation (ITAVI)

fish, while still being more than sufficient for the development of nitrifying bacteria and plant growth (Rakocy et al., 2006). At the same time, CO_2 levels must be kept below 15 mg/l to prevent fish suffocation. Sufficient degassing must therefore be ensured.

15.2.4. Electrical conductivity (EC)

Electro-conductivity (EC) measures water's ability to conduct electric current. It is expressed in µS/cm. All the ions (electrically charged molecules) in water play a part in this: they are also called "mineral salts", and are all the nutrients present in water, which may or may not be involved in plant nutrition. Pure water has no electro-conductivity. Conductivity can also be measured by the dissolved solids content (DSC) expressed in mg/l or ppm (parts per million). The conversion factor is as follows: TDS (ppm) = 0.67× conductivity (µS/cm) (formula valid up to 2,000 µS/cm). Too high a TDS or conductivity can lead to dehydration of the plants.

This parameter is important to monitor in hydroponics as it is directly proportional to the quantity of mineral salts dissolved in the solution, and reflects the force of osmotic pressure (Trejo-Tellez et al., 2012). Osmotic pressure induces a nutrient concentration differential between the solution and the plant tissues, which promotes the diffusion of dissolved substances through the plant membranes (suction phenomenon).

15.2.4.1. AQUACULTURE COMPARTMENT. Conductivity in aquaponics is not a risk factor for freshwater fish. A risk linked to this parameter occurs for a water salinity greater than 5 ‰, which corresponds at 10°C to a conductivity of 6,000 µS/cm. Similarly at 20°C, this corresponds to a conductivity of 8,000 µS/cm. These values will never be reached whether in aquaponics or hydroponics.

15.2.4.2. PLANT COMPARTMENT. It is very difficult to predict the level of conductivity that water will have in a system, as there are so many environmental variables to take into account: frequency of feeding, digestibility of feed, rate of retention of solids in the mechanical filter, mineral input from renewal water, water hardness, etc.

In a conventional hydroponic solution, the recommended range is 1,500 to 2,500 µS/cm up to 3,000 µS/cm or more for certain plants such as tomatoes, which tolerate high salinity. An aquaponic system rarely produces such high conductivities: in a correctly sized, fairly intensive system, low levels of 600 to 1,200 µS/cm are generally achieved, although it is possible to exceed this indicative threshold with a completely closed fish circuit. Contrary to all expectations, despite this significant difference, productivity in aquaponics can be just as good as in hydroponics, which once again calls into question certain knowledge taken for granted: aquaponics is a new paradigm for soil-less cultivation.

According to some authors, the difference is partly due to the fact that nutrients are generated continuously through livestock effluents (raft and NFT techniques), and not distributed to the the roots by watering at fixed intervals as in conventional hydroponics (Rakocy et al., 2006). However, work carried out as part of the APIVA® project has shown that it is entirely possible to grow plants successfully using tidal tables or drip irrigation. The fact that aquaponics can enable satisfactory plant growth at such low conductivity levels (generally < 1200 µS/cm) is a real curiosity for soilless culture specialists. According to other authors, conductivity is not really representative of the true nutrient contribution of fish effluent to plants. At most, this parameter gives an indication of the quantity of nutrients released by the fish in mineral form, but no indication of the forms linked to organic particles (Goddek, 2015). Particles that are too fine to be retained by mechanical filtration techniques tend to accumulate over time in the system, particularly at the bottom of the culture surfaces, which play a decanting role. These particles can be mineralised by heterotrophic bacteria and other micro- or macro-organisms, resulting in forms that can be assimilated by plants, even at pH levels that are too high for plant cultivation. Some studies suggest that plants are also capable of assimilating non-mineral forms of nitrogen, such as amino acids. Ghosh and Burris (1950), for example, found that tomatoes can use alanine, glutamic acid, histidine and leucine just as efficiently as inorganic sources of nitrogen. It is now increasingly suspected that plants are able to absorb other amino acids, vitamins and antibiotics produced in the rhizosphere by fungal and bacterial activity, although the mechanisms involved are not fully understood.

Therefore, measuring conductivity is only of limited relevance in aquaponics if you want to measure the nutrient potential of the water. However, it is still useful for monitoring changes in the quantity of mineral nutrients in the water. It is possible, for example, to make correlations between the quantity of nitrates and conductivity in order to predict the concentration of nitrates in the water on the basis of conductivity.

15.2.5. Total water hardness and full alkalimetric titre

The total hardness of water - or hydrotimetric titre (TH, also known as GH) - is an indicator of the mineralisation of water, linked essentially to the calcium $Ca^{(2+)}$ and magnesium Mg^{2+} ions in solution. It is expressed in French degrees (°f): 1 °f corresponds to a concentration of 4 mg/l of calcium and 2.4 mg/l of magnesium. In everyday life, hardness is reflected in the formation of limescale deposits in pipes. Ideal hardness is between 15 and 25°f. Below 15°f, water is considered "soft", above 30°f, it is considered "hard".

Total alcalimetric strength (TAC) measures the ion concentrations of bicarbonate (HCO_3^-), carbonate (CO_3^{2-}) and hydroxide (OH^-). It is also expressed in °f, as the equivalent concentration of calcium carbonate ($CaCO_3$). 1 °f corresponds to 10 mg/l of calcium carbonate ($CaCO_3$). The more there is, the more alkaline the water. The less it contains, the more acidic it is. The alkaline and acid reserves in the water make up its buffering capacity, i.e. the degree to which its pH varies when an acid or base is added.

The advantage of 'hard' alkaline water is that it has a greater pH buffering effect than 'soft' water. The disadvantage of water that is too hard is the possibility of precipitation of calcium and magnesium in the form of calcium and/or magnesium carbonates, or calcium phosphates, which are insoluble compounds that can cause phosphorus deficiencies in plants due to reduced bioavailability to the roots. A hardness of 15 to 25°f is suitable for the vast majority of freshwater fish, and is an adequate range for plants and nitrifying bacteria.

15.2.6. Redox potential

The oxidation-reduction potential or redox potential is expressed in mV and is used to qualify an aqueous solution and classify it according to whether it is oxidising (presence of oxygen) or reducing (lack of oxygen).

Oxidation-reduction reactions play an important role in the behaviour of different elements in the environment, particularly in the transformation of biological compounds (Matia *et al.*, 1991). Redox potential is the main parameter that controls the dynamics of these reactions.

Generally speaking, it is accepted that a low redox potential reflects water loaded with particulate nutrients, and that a high redox potential is a good indicator of the cleanliness and oxidising power of the water. This potential affects the rate of mineralisation of organic compounds present in aquaculture water. A high positive redox potential implies a high rate of mineralisation of suspended matter (Savidov, 2014) and is a good indicator of good quality, well-oxygenated water. A negative oxidation-reduction potential indicates a lack of oxygen and may be indicative of anaerobic activity of the bacteria present in the medium on the suspended organic particles, and therefore of a 'reducing' activity that is generally accompanied by the release of ammonium and nitrites or even hydrogen sulphide H_2S.

Thus, redox potential is closely linked to the quantity of oxygen present in the environment and to the quantity of suspended organic matter and algae. It is also linked to the pH and salinity of the water, in a way that is still poorly described in the scientific literature.

In aquaculture as in hydroponics, it is accepted that an adequate redox potential is between +100 and +300 mV. The most important thing is to have a certain stability of this value over time: sudden changes in the redox potential reflect a chemical imbalance in the system that is not always immediately apparent from the other physico-chemical parameters. This parameter is not easy to interpret.

15.3. Organic load, suspended solids rate

If the solids discharged by the fish are not filtered and removed, they will accumulate and could potentially become a risk factor if they are degraded in the water by oxygen-consuming heterotrophic bacteria and vectors of ammonia discharges, which could ultimately have an impact on the development of plants and, above all, on the health of the fish (Pambrun, 2005; Graber, 2009; Trejo-Tellez *et al.*, 2012; Yildiz, 2017). Figure 3-10 illustrates the existence of competition for oxygen and nutrition between autotrophic and heterotrophic bacteria on bacterial attachment media in the biological filter.

It is common in the literature to find small-scale aquaponic systems operating on the model of inert substrates (*Media Filled Growbed* - MFG), in which the plant culture surface is both the support for the plants and the filter medium for managing solid and dissolved particles. This can work on a non-commercial scale, and at low fish stocking densities of around 5 to 10 kg/m³. Beyond that, because of the inherent difficulty in maintaining these culture media, the foreseeable consequence in the medium term is their clogging, a reduction in the efficiency of mechanical filtration and the appearance of anaerobic zones favouring the emergence of undesirable biological phenomena (release of ammonia and hydrogen sulphide).

A system that aims to be commercial and large-scale should rather rely on a mechanical filtration process, the efficiency of which is proven and not directly dependent on biological phenomena, which can change quite randomly over time. As a guideline, it is recommended to ensure a suspended solids (SS) concentration of less than 30 mg/l (Rakocy *et al.*, 2006) in rearing tanks, with wide variations in tolerance depending on the species of fish.

Biochemical demand and chemical oxygen demand (BOD and COD) are two parameters that reflect the amount of organic matter in the system and the quality of the water: these two parameters should be kept as low as possible.

16. Conclusion

The main elements for sizing aquaponics systems are relatively well established. However, each system is unique in terms of design, technical choices and cultivation strategies. Several factors influence the balance of an aquaponics system:

– the pair of fish and plant species under consideration (intrinsic zootechnical and phytotechnical data) and the production strategy;

– the quantity and quality of the fish feed (protein concentration, digestibility, nature of the plant or animal raw materials);

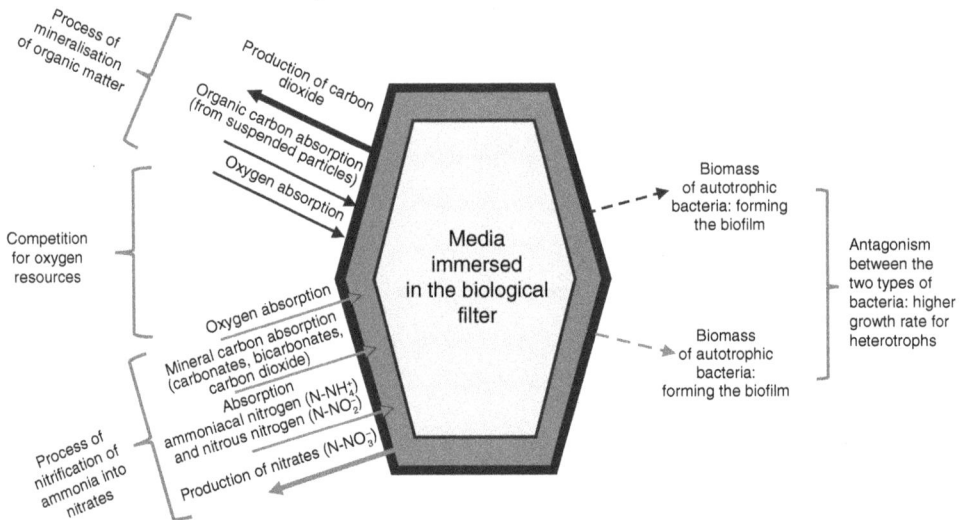

Figure 3-10. Illustration of the existence of autotrophic/heterotrophic bacterial competition (Pierre Foucard, ITAVI)

- the type of hydroponic system (*rafts*, NFT, systems on inert substrate, in pots, tidal tables, etc.);
- the system's hydraulic parameters (flow rates, retention times in the biofilter and in the plant compartment, hourly renewal rate of the fish tanks and plant culture structures, etc.);
- the quality of new water and variations in the physico-chemical parameters of the water.

Table 3-4 summarises the values of the main parameters that are important to know and monitor in aquaponics. These values are compromises to satisfy all the organisms raised or cultivated together in aquaponics (nitrifying bacteria, fish and plants). It is neither advisable nor feasible to measure all these parameters on a daily basis.

Table 3-4. Guide to the main "compromise" values to target and monitor regularly to meet the specific needs of bacteria/fish/plants in aquaponics. (Pierre Foucard)

Parameter	Compromise in aquaponics
Temperature	[15–25 °C]
pH	[6.5–7.5]
Dissolved $O_{(2)}$	> 5 mg/l
Dissolved CO_2	< 20 mg/l
MY	< 30 mg/l
TAN (at T 15°C; pH 7)	< 2 mg/l
Ammonia N-NH_3	< 0.01 mg/l
Nitrite N-NO_2^-	< 0.3 mg/l
Nitrates N-NO_3^-	< 250 mg/l
Conductivity	[600-1,200] µS/cm
Alkalinity	> 100 mg/l $CaCO_3$
Water hardness	[15–25] °F
Redox potential	[100 to 300] mV

The most important parameters to monitor on a daily basis are temperature, dissolved oxygen and pH, so it's essential to buy fixed or portable probes.

Ammonia and nitrite levels should be monitored daily during the biofilter seeding phase and while the fish are acclimatising to their new living environment: this can be done using professional-quality test strips, or spectrophotometers for greater accuracy but at a higher cost. Thereafter, ammonia and nitrite levels should not reach excessive values as long as the biological filter is correctly sized, and monitoring can then be limited to a weekly basis.

Similarly, TSS levels need to be monitored, but are not normally a problem if the mechanical filter and the opening rate of the system are correctly dimensioned to avoid excessive accumulation. The level of CO_2 should not exceed 20 mg/l, and to achieve this you need to maximise air/water exchanges: a sufficiently aerated biofilter (bubbling) ensures good degassing of carbon dioxide.

Hardness and alkalinity are interesting parameters to measure in new water, even before installing an aquaponic greenhouse, but do not require daily monitoring afterwards.

Conductivity and nitrate levels should be monitored on a fortnightly basis to observe changes in the mineral concentration of the nutrient solution. It is also worth measuring phosphorus and potassium concentrations, and even other macro- and micro-elements, to obtain a nutrient profile of the water and compare these values with the needs of the plants, so as to anticipate any deficiencies in a particular element and take appropriate measures if necessary.

4

Regulatory, societal and economic challenges

The technical aspects of aquaponics are fairly well documented, thanks to the various research projects carried out around the world over the last thirty years. However, the development of aquaponics is still limited by the legal, societal and economic issues that remain.

The question now arises as to the regulatory framework(s) on which aquaponics could depend: what status should operators have between fish farming, market gardening and horticulture? And to what extent could a producer be authorised or not to sell products (depending on their 'status') from these systems, which are not very widespread and not very well known by the public and administrative bodies?

What's more, the question of the social acceptability of this method has been assessed far too little to date and remains a real issue: are consumers prepared to accept this production system and consume its products?

For the moment, there seems to be a consensus that aquaponics is "good for the environment", but there is no real scientific basis for this claim. The best tool currently available for assessing the environmental impact of an activity is Life Cycle Assessment (LCA), a standardised evaluation method that enables a multi-criteria, multi-stage environmental assessment of a product or process to be carried out throughout its life cycle, from the extraction of the raw materials required for its manufacture, through its distribution, marketing and use phases, to its end-of-life: Very few studies have really developed this

line of research for aquaponics, the interest being to compare the environmental impact of this practice with other systems for the same amount of service provided.

Finally, many consultancies are currently positioning themselves to offer the design and construction of aquaponics systems on a commercial scale, but the conditions for the profitability of this method of production have not been fully recognised or documented: no replicable model or model farm exists. In short, are we really in a position today to say that aquaponics is a good idea from a purely economic point of view?

17. Regulatory challenges

Regulation is a crucial aspect that is all too often neglected in the early stages of a project to set up a fish farm, as the body of French regulations is vast, complex and often not very explicit for activities such as fish farming, and even less so for new activities such as aquaponics. Support, awareness-raising and expertise are needed for project developers.

Aquaponics involves root irrigation of plants with water that could be considered as 'wastewater', while at the same time involving two very different types of production - fish and plants - each with its own constraints and regulatory issues (environmental, health, etc.) that need to be understood and taken into account.

©2026 CAB International. *Aquaponics* (eds Pierre Foucard and Aurélien Tocqueville) DOI: 10.1079/9781836991441.0004

Like aquaculturists, aquaponiculturists use a shared primary resource (water) and generate effluents (Hoevenaars, 2018), and their activities are subject to a large number of regulations at different levels. The development of aquaponics is still at an embryonic stage and, moreover, is often carried out in urban or peri-urban areas where regulatory approaches may differ from activities in rural areas.

17.1. General regulations governing the use of water

The European Water Framework Directive (WFD) 2000/60/EC introduced water pricing based on the "polluter pays" principle to encourage users to use water resources efficiently. In addition, one of the founding principles of this directive is *to* eventually achieve 'good status' for aquatic environments.

This Community text was transcribed into French law in 2004. The 2006 law on water and aquatic environments (LEMA) (law no. 2006-1772 of 30 December 2006) provided the framework and tools needed to achieve the objectives set by the WFD. This law strengthens State action (water policing), and sets a "nitrogen" charge for farms discharging surpluses calculated on the basis of the annual nitrogen input-output balance, as well as charges for water consumption - surface or groundwater - (which does not apply to aquaculture), and for altering the water regime. It also sets a non-domestic pollution charge payable to the water agencies, particularly for fish farming.

The Water Act led to the introduction of the SDAGE (Schéma d'aménagement et de gestion des eaux - Water Development and Management Scheme), through the Water Agencies' basin committees, and the SAGE (Schéma d'aménagement et de gestions des eaux - Water Development and Management Scheme), on a catchment or sub-catchment scale, through the Commissions locales de l'eau (CLE - Local Water Commissions). All users linked to aquatic environments - including fish farmers - are present and must be represented in these decision-making bodies on specific rules that may apply locally. The same will undoubtedly apply in the future to aquaponics producers, depending on their link with an aquatic environment, for the purposes of drawing water from and/or discharging it into that environment.

17.2. General regulations applying to fish farming

17.2.1. Fish farm status?

Fish farming is an activity defined in article L 431-6 of the French Environment Code: "A fish farm is, within the meaning of Title I of Book II and Title III of Book IV, a farm whose purpose is the rearing of fish for consumption, restocking, ornamental, experimental or scientific purposes, as well as for tourism. In the latter case, fish may be caught using lines in bodies of water.

It is therefore appropriate to consider the application of this definition and the status applicable to an aquaponics site, which may produce fish or other aquatic organisms, the purposes of which are specified in this definition, as well as plant production, which is usually much greater in volume than aquaculture production.

17.2.2. Authorisation and declaration schemes for fish farms

Fish farms currently come under two different regimes, depending on the production capacity of the sites.

17.2.2.1. ICPE: INSTALLATIONS CLASSIFIED FOR ENVIRONMENTAL PROTECTION. NB: seawater fish farms are exclusively classified as ICPEs. Furthermore, depending on where they are located, they may be subject to an additional procedure: an "application for authorisation to operate marine cultures, in order to obtain a concession on the public maritime domain".

This classification is based on the concept of production capacity. This is not defined in any specific regulations or instructions. However, it is generally accepted that it is based on biomass production at a site and therefore on a variation in annual stock. Production capacity can be expressed as follows: *production capacity = ([final stock− initial stock] - purchases + sales)*

17.2.2.2. IOTA: INSTALLATIONS, WORKS AND ACTIVITIES. Fish farms are also covered by the "Water" nomenclature. In particular, those that do not fall under the previous ICPE category are covered by the heading: 3.2.7.0. "Freshwater fish farms mentioned in article L 431-6 (D)" (under heading "III. Impacts on the aquatic environment or on public safety" of the Water nomenclature), with a production capacity of less than twenty tonnes per year.

However, project promoters will need to check, depending on their situation, whether or not the other headings in the Water nomenclatures (Tables 4-1, 4-2 and 4-3) apply.

It is therefore necessary to carry out a full assessment of the body of IOTA/ICPE regulations prior to submitting an application, depending on the characteristics of the project. A site that is subject to declaration on the basis of its production capacity may be switched to

authorisation by the application of just one other heading subject to authorisation.

17.2.3. The environmental authorisation procedure

Since 1er March 2017, an IOTA or ICPE dossier has had to follow the same procedure before reaching a prefectoral decree: the "environmental authorisation procedure".

This procedure applies to new authorisations, renewals of existing authorisations or amendments to existing orders.

Table 4-1. Heading 2130 of the ICPE nomenclature: published on 27 July 2006, effective 1st October 2006. (Ministry of the Environment, Energy and the Sea)

N°	Description of heading	A, D, S, C[1]	Radius[2]
2130	Fish farms		
	1. Freshwater fish farms (excluding stocked ponds, where rearing is extensive, without feeding or with exceptional feeding), with a production capacity of more than 20 t/year	A	3
	2. Seawater fish farms, with a production capacity of:		
	a) more than 20 t/year	A	3
	b) more than 5 t/year but not more than 20 t/year	D	

[1]A: authorisation, D: declaration, S: public utility easement, C: periodic inspection under article L. 512-11 of the French Environment Code.
[2]Display radius in kilometres.

Table 4-2. IOTA nomenclature (Ministry of Ecological Transition and Solidarity)

IOTA category	Structures or impacts
1.1.2.0.	Permanent or temporary withdrawals from a borehole; the total volume withdrawn per year being: – greater than or equal to 200,000 m³ (authorisation); – between 10,000m³ and 200,000 m³ (declaration)
1.2.1.0.	Withdrawals, including by diversion from a watercourse; the volume withdrawn being: – greater than 1,000m3 per hour or 5% of the flow of the watercourse (authorisation); – between 400 and 1,000 m³ per hour or between 2 and 5% of flow (declaration)
3.3.1.0.	The draining, impounding, sealing or filling of wetlands or marshes, the area drained or impounded being: – greater than or equal to 1 ha (authorisation); – greater than 0.1 ha but less than 1 ha (declaration)

Table 4-3. ICPE nomenclature (Ministry of Ecological Transition and Solidarity)

ICPE category	Designation
4725: Oxygen	The quantity likely to be present in the installation being: – equal to or greater than 200 t (authorisation); – greater than or equal to 2 t but less than 200 t (declaration) *Low threshold quantity within the meaning of article R. 511-10: 200 t.* *High threshold quantity within the meaning of article R. 511-10: 2,000 t.*

This recent reform should reduce the time taken to issue a decree to ten months.

Aquaponics projects will have to follow this procedure. Depending on the configuration and issues at stake, and in particular the IOTA and/or ICPE headings mentioned above, a project may be subject to an environmental assessment with a dossier containing an impact study, and may be submitted to a public enquiry. In addition, the environmental authorisation procedure is linked to the town planning procedures (particularly if a building permit is required, Table 4-4) and the public enquiry is then a single one if it is required by both decisions.

Project promoters should contact the instructive services that can guide them through the process: DDT, DDTM, DDPP, DDCSPP, etc.

17.2.4. Operating requirements for a fish farm

Defined in decrees dating from 1st April 2008 for freshwater fish farming, the operating rules by ministerial decree do not exist for marine fish farming:

– The decree of 1st April 2008 (OJ of 12/04/2008) defines the technical rules to be met by freshwater fish farms subject to authorisation under Book V of the Environment Code (ICPE);
– The Order of 1 April 2008 sets out the general requirements applicable to installations,

works or activities subject to declaration pursuant to Articles L. 214-1 to L. 214-6 of the Environment Code and falling under heading 3.2.7.0 of the nomenclature appended to the table in Article R. 214-1 of the Environment Code (freshwater fish farms mentioned in Article L. 431-6) and repealing the Order of 14 June 2000.

In freshwater fish farming, these decrees specify various operating rules that must be incorporated into the design phase of projects: noise, integration into the landscape, restoration, spreading, monitoring, controls, self-monitoring, standards for discharges into the aquatic environment, and so on.

All of these requirements have been developed in relation to the situations most frequently encountered in freshwater, with sites operating directly close to a watercourse by diversion from it. It would therefore appear to be very delicate, and incoherent for certain prescriptions, to envisage an application to aquaponics sites, based on the operation of the fish farming part in a recirculated circuit as widely described in this book.

Certain requirements may already be problematic for aquaponics projects, such as those concerning the location of the site. This issue could become increasingly important, particularly in view of the number of aquaponics projects in urban environments. Fish farming facilities must be installed:

– at least 100 metres from the dwellings of third parties (with the exception of dwellings occupied by personnel of the installation and rural gîtes enjoyed by the operator) or premises usually occupied by third parties, stadiums or approved campsites (with the exception of farm campsites) as well as areas designated for housing by town planning documents enforceable against third parties;
– at least 3 kilometres upstream or downstream of an existing fish farm on the same watercourse (this distance is measured immediately upstream of the water intake or immediately downstream of the discharge, along the axis of the watercourse);
– within a radius of at least 1 kilometre of a fish farm located in the same catchment area.

Table 4-4. Planning formalities and description of works to be built, in the case of the installation of frames or greenhouses (from *Installations Agricoles. Guide technique pour l'instruction des autorisations d'urbanisme. Bretagne*, May 2016)

Frames or greenhouses

If height ≤ 1.80 m (regardless of surface area)
Principle: no planning formalities required
R. 421-2(e) of the town planning code
If 1.80 m < height < 4 m and floor area ≤ 2000 m²
on the same plot[1]
Principle: prior declaration
R. 421-9(g) of the town planning code
If height ≥ 4 m
Principle: planning permission
If height > 1.80 m and floor area > 2000 m² on the
same plot[1]
Principle: planning permission
R. 421-1 of the town planning code

There are, however, exemptions to this siting rule, which is not specific to fish farming but applies to all types of farming. It would therefore seem realistic to apply to the prefecture for an exemption, given the specific nature of the activity, and to justify that the project does not present any nuisance and that everything is planned to limit/manage the impact, or even not to impact third parties through noise, odours, visual disturbance, etc. This requires time and costs to develop a complete file and to have it studied. This type of exemption has already been granted for other farms.

17.2.5. Discharge of fish effluent in urban areas ?

In an urban environment, it is far from easy to manage the solid and liquid effluent from an aquaponics farm, and this issue is all too often overlooked when the project is being considered. Fish farm water and sludge cannot, *a priori*, be discharged into the urban wastewater network. Only domestic wastewater and industrial wastewater are authorised, as defined in the special discharge agreements signed between the sanitation department and industrial establishments when applications are made to connect to the public network.

Each prefecture defines a policy on discharges into collective sewerage systems. In general, it is forbidden to discharge any type of effluent from agricultural livestock farming (liquid manure, slurry, etc.): the question of the definition and status of water from an aquaculture farming system remains, however.

It is therefore possible to submit a request or apply for an exemption, depending on the specific regulations in force in the municipality: annual volume of effluent to be treated, compliance with maximum permissible concentrations for various parameters (COD, BOD5, SS, total nitrogen, total phosphorus, etc.). A request for authorisation must then be submitted to the "wastewater and sanitation" department, so that the discharge can be investigated prior to any construction project. Once again, it seems realistic to achieve this, but patience is required.

17.2.6. What fish species are authorised in recirculating systems?

The decree of 17 December 1985 sets out the list of species of fish, crustaceans and frogs represented in France (carp, perch, pike-perch, catfish, trout, tench, etc.) and therefore authorised for introduction for aquaculture production.

In addition to the question of authorised species, we need to consider the "authorisable species", which do not appear on the previous list of the 1985 decree. These non-represented species, which can be introduced after submitting a specific application to the prefecture, are included in the Order of 20 March 2013 and concern the love carp *(Ctenopharyngodon idella)* and various species of sturgeon. The latter, in a specific and unique context for caviar production purposes, are included on a list in the Order of 23 February 2007 "laying down the conditions for authorising the introduction of sturgeons and the authorisation procedure for establishments packaging or repackaging caviar for export, re-export or intra-Community trade".

Following on from the Order of 20 March 2013, the Order of 6 August 2013 sets out, in application of article R. 432-6 of the Environment Code, the form and content of applications for the authorisations provided for in articles L. 432-10 and L. 436-9 of the Environment Code: "any application for authorisation to introduce an unrepresented species, other than for scientific purposes, shall be addressed to the prefect of the département where the introduction is planned" as long as it does not concern aquatic animals that are on the list mentioned in article R. 432-5 of the Environment Code, which are likely to cause biological imbalances, and whose introduction (possession and transport) is prohibited: namely the catfish *(Ictalurus melas)*, the sun perch *(Lepomis gibbosus)*, as well as various species of crustaceans and frogs.

In special cases (ornamental species), it is advisable to consult the lists of species, breeds or varieties of domestic or non-domestic animals for which specific rules apply (certificate of competence, etc.). All other species not listed may potentially be authorised subject to a prefectoral decision, with particular reference to Decree no. 2017-595 of 21 April 2017 on the control and management of the introduction and spread of certain animal and plant species, in order to limit the spread of invasive alien species following the 2016 Biodiversity Act, even if the outcome of such procedures seems unlikely or subject to severe constraints.

However, at European level, Regulation (EC) No 708/2007 (amended by Regulation (EC) No 304/2011 of the European Parliament and of the Council of 9 March 2011) sets out the conditions for "the use of alien and locally absent species in aquaculture", establishing a framework governing aquaculture practices in order to assess and minimise the potential impact of these species on aquatic habitats, thereby contributing to the sustainable development of the sector. Annex IV of this regulation sets out the species recognised as "present in France", although they are not included in the above-mentioned lists.

This regulation applies without prejudice to the provisions laid down in Regulation (EU) 1143/2014 of the European Parliament and of the Council of 22 October 2014 on the prevention and management of the introduction and spread of invasive alien species, Articles L. 411-5 to L. 411-7 and L. 432-10 of the Environment Code and the implementing texts adopted for the fight against invasive alien species.

The regulation, which is directly applicable in France, stipulates that these species must be subject to expert appraisal according to the principles of risk analysis: an application for an introduction permit is required from the competent authority designated by the Member State. This authority organises a scientific assessment to evaluate the risks associated with the introduction, as well as the mitigation and monitoring measures and emergency plans needed to achieve a low risk. A period of 6 months is required for the assessment. The Council and/or the other Member States must be notified if the application is likely to affect them.

It should be noted that so-called "closed" aquaculture facilities are considered specifically: the aquaculture of exotic or "locally absent" species is made possible, without the need for an introduction permit. Closed aquaculture facilities are specifically defined as "land-based facilities" where "aquaculture is carried out in an aquatic environment involving recirculation of water" and where "discharges have no connection whatsoever with open water prior to screening and filtration or percolation and treatment to prevent the release of solid waste into the aquatic environment and any escape from the facility of farmed and non-target species likely to survive and subsequently reproduce" and which:

1. prevents losses of farmed or non-target species and other biological material, including pathogens, due to factors such as predators (e.g. birds) and flooding (e.g. the facility must be located at a safe distance from open water following an appropriate assessment by the competent authorities);
2. prevent, by reasonable means, losses of farmed or non-target species and other biological material, including pathogens, due to theft and vandalism;
3. ensures the proper disposal of dead organisms.

In practice, it remains extremely difficult to obtain authorisation for the introduction of species that are not represented in the French context, even in recirculated systems. Nevertheless, work and discussions (definitions, inventories of sites, administrative management, etc.) are underway, particularly on the application of the EU regulation, in connection with the development of recirculating aquaculture projects, such as aquaponics.

18. Regulations on irrigating crops with wastewater

Livestock effluents, sewage sludge, organic urban waste and industrial effluents - known as waste fertilising materials (WFM) - are sources of fertilising elements and organic matter for fertilising or amending agricultural or forestry soils. Spreading livestock manure and other residual fertilisers provides the mineral elements (nitrogen, phosphorus, potassium - N, P, K) needed for plant nutrition. It also provides organic matter with amending value, helping to improve soil properties. While the regulations governing the spreading of agricultural sludge are well documented, it has proved difficult to find regulatory data on the irrigation of crops with 'waste water' from fish farms. The reason for this is that fish farm effluent is usually discharged directly into watercourses and not reused for food production.

On the basis of an opinion issued by the French Food Safety Agency (Afssa, 2008) "on the risk assessment of effluents from establishments processing category 1, 2 or 3 by-products for reuse in the irrigation of crops intended for human or animal consumption", the relevant ministries (health, environment and agriculture) published the Order of 2 August 2010 in the *Official Journal* "relating to the use of water from urban wastewater treatment for the irrigation of crops or green spaces". The aim of these regulations is to ensure the protection of public health, animal health and the environment, as well as the safety of agricultural production. According to article 5 of this decree, the irrigation of crops and green spaces is prohibited "using treated wastewater from wastewater treatment plants connected to an establishment for the collection, storage, handling or treatment of category 1 or 2 animal by-products within the meaning of European regulation 1069/2009 and subject to the regulations governing classified installations [...] with the exception of cases where the water is heat-treated at 133°C for 20 minutes under a pressure of 3 bars before being discharged into the collection system" and using "raw wastewater".

For various reasons, the Order of 2 August 2010 seems irrelevant to the specific case of aquaponics. Its scope is limited to the irrigation of crops or green spaces (all irrigation techniques concerned) "with treated wastewater from wastewater treatment plants mentioned in II of article L. 2224-8 of the General Local Authorities Code and from non-collective sanitation installations mentioned in III of article L. 2224-8 of the General Local Authorities Code, and whose gross organic pollution load is greater than 1.2 kg of five-day biological oxygen demand (BOD5) per day".

According to European regulation no. 1069/2009 "laying down health rules concerning animal by-products and derived products not intended for human consumption, animal by-products and derived products fall into categories 1, 2 and 3 as set out in the Order of 2 August 2010", categories described and explained in regulation 1774/2002. However, the scope of this article does not include the following animal by-products: "excrement and urine other than liquid manure, and non-mineralised guano". Fish farming effluents are clearly not guano, nor manure, according to the definition detailed in the article of the law, which defines manure as "any excrement and/or urine of farmed animals other than fish, with or without litter".

To sum up, the Ministerial Order of 2 August 2010 is the main regulatory reference whose scope could come close to the case of crop irrigation using wastewater from aquaculture.

However, aquaculture effluents have nothing in common with urban wastewater in physico-chemical and microbiological terms; moreover, fish farming effluents do not fall within the definitions of category 1, 2 or 3 waste cited in this decree and defined in European regulation no. 1069/2009. In the absence of an appropriate legal framework, an activity such as aquaponics should at least be able to justify the absence of health risks if the plants produced are intended for human consumption. It is important to show that the irrigation technique used does not involve contact of the culture water with the edible parts of the plants produced, and that regular water quality checks are carried out in accordance with Regulation (EC) No 852/2004 of the European Parliament and of the Council of 29 April 2004 on the hygiene of foodstuffs. Sprinkling water on the leaves should be avoided in aquaponics to prevent any health problems.

Although not adapted to the aquaponics context in terms of effluent categorisation, the ministerial decree of 2 August 2010 defines four categories of use for treated wastewater. The category with the most stringent associated standards (category A) covers the irrigation of unprocessed or processed market garden crops and the watering of green spaces open to the general public. The physico-chemical water quality parameters expected for category A seem realistically achievable for fish farm effluent. It could therefore be proposed to measure the microbiological and physico-chemical quality of the water used for irrigation in aquaponics, at the outlet of the "aquaculture" compartment and the "biofilter" compartment of the system, in order to validate the possibility of classifying it in use category A, which must meet the health criteria defined in Table 4-5.

19. Health regulations

19.1. Regulations relating to the sanitary aspects of animals and water

19.1.1. Health regulations for fish farms

The specific "health" regulatory corpus for fish farming is just as vast as the previous regulatory themes, and needs to be understood precisely for aquaponics projects, as the stakes are high for

Table 4-5. Parameters to be respected in order to be in category A for use of treated wastewater, according to appendix 1 of the ministerial order of 2 August 2010. (Légifrance)

Parameters	Criteria for use category A (for market gardening)
Suspended solids (mg/l)	< 15
Chemical oxygen demand (mg/l)	> 60
Eschersischia coli (CFU/100 ml)	< 250
Faecal enterococci (log abatement)	≥ 4
F-specific RNA phages (log knockdown)	≥ 4
Spores and sulphite-reducing anaerobic bacteria (log abatement)	≥ 4

the industries concerned. France is committed to a specific action plan: "Fish Health Plan 2020".

The main texts concerned are as follows:

–   Directive 2006/88/EC of 24 October 2006 on animal health requirements for aquaculture animals and products thereof, and on the prevention and control of certain diseases in aquatic animals;
–   Commission Decision of 31 October 2008 implementing Council Directive 2006/88/EC as regards surveillance and eradication programmes and disease-free status of Member States, zones and compartments (notified under document number C[2008] 6264) (Text with EEA relevance);
–   arrêté du 4 novembre 2008 relatif aux conditions de police sanitaire applicables aux animaux et aux produits d'aquaculture et relatif à la prévention de certaines maladies chez les animaux aquatiques et aux mesures de lutte contre ces maladies.

Various health hazards are defined and classified into different categories. First category hazards for fish are defined in the Order of 29 July 2013. The corresponding technical measures are defined in the Order of 4 November 2008. These are listed here:

–   infectious haematopoietic necrosis (IHN) and viral haemorrhagic septicaemia (VHS). These two diseases are rhabdoviroses present in

France mainly in salmonids, and only occur at water temperatures below 14°C;
–   Carp herpes virus (CVH). This disease, which is present in France, mainly affects carp. As it is a herpes virus, symptoms only appear at water temperatures above 18°C;
–   Infectious salmon anaemia (ISA). France is free of ISA, but the disease is present in Europe, particularly in Norway. The disease affects salmon, rainbow trout and brown trout. The category 1 hazard is the genotype deleted in the highly polymorphic region (HPR) of the isavirus genus (ISAV), which is responsible for the symptomatic form of the disease;
–   Epizootic haematopoietic necrosis (EHN). France is free from EHN, a disease classified as exotic by Directive 2006/88/EC because it has not been observed in the European Union to date.

NB: it is important to remember that fish diseases are not zoonoses and cannot be transmitted to humans.

France is committed to a national eradication and surveillance plan (PNES) for viral haemorrhagic septicaemia (VHS) and infectious haemoatopoietic necrosis (IHN). All facilities are involved, especially those holding susceptible species and vectors[7] of these diseases. The ultimate aim of this approach is to make the whole of France free of these diseases.

Thus, in order to be introduced into zones or sites recognised as officially free, or implementing qualification or eradication programmes with regard to one or more of the endemic diseases regulated on French territory or in a Member State, aquaculture animals must themselves come from an aquaculture zone or compartment free from the disease(s) in question.

Sites may only exchange animals with other sites of equivalent or higher health status: wild aquatic animals belonging to susceptible species are, if necessary, placed in quarantine before being introduced into a free zone or compartment in accordance with Decision 2008/946/EC (the health conditions relating to exchanges between Member States and the model health certificates to be used are specified in Regulation [EC] 1251/2008).

Health qualification leading to "free" status for one or more endemic diseases for an aquaculture zone or compartment is requested by the operator on the basis of a 2-year or 4-year qualification programme. Once completed, the programme is declared to the European Commission, to which it is submitted for approval.

In addition, the marketing of aquaculture animals, whether for profit or not, is an activity subject to animal health approval: AZS. As an exception to the general principle of animal health approval, certain aquaculture establishments or farms are only subject to registration. Animal movements must not jeopardise the health status of animals at destination and transit sites. The placing on the market of aquaculture animals can therefore only be carried out by aquaculture farms with animal health approval.

Animal health approval provides a framework for placing aquaculture animals on the market. It includes the registration of animal movements, the declaration of unexplained increases in mortality, a risk analysis of aquaculture farms, the application of good aquaculture health practices and the implementation of an animal health surveillance plan as set out in the Order of 8 June 2006. The application should be submitted as soon as possible, as approval must be obtained before the animals are placed on the market; it is therefore advisable, including for aquaponics projects, to plan this process in parallel with that for setting up the site. To apply for approval, a specific form must be sent, together with the required documents, to the DD(CS)PP (Departmental Directorate for Social Cohesion and Protection of Populations) where the activity is located. Before submitting the application, it is advisable to contact a veterinarian specialising in aquaculture and the DD(CS)PP in order to define the necessary elements and carry out a risk analysis taking into account the specific features of the site and the activity.

Finally, all health procedures (care, treatment, etc.) on aquatic animals must be carried out under the responsibility of a veterinarian, who must be designated for the production site.

19.1.2. Regulations relating to microbiological parameters on plants

Water can be a vector for many micro-organisms, including pathogenic strains of bacteria such as *Escherichia coli, Salmonella, Vibrio cholerae and Shigella*, and microscopic parasites such as *Cryptosporidium parvum, Giardia lamblia, Cyclospora cayetanensis and Toxoplasma gondii*. Even small amounts of contamination with some of these organisms can cause food-borne illness in humans.

Regulation (EC) No 2073/2005 lays down the microbiological criteria for certain micro-organisms and the implementing rules that food business operators must observe when implementing the general and specific hygiene measures referred to in Article 4 of Regulation (EC) No 852/2004. The competent authority shall verify compliance with the rules and criteria laid down in this Regulation in accordance with Regulation (EC) No 852/2004, without prejudice to its right to undertake further sampling and analysis to detect and measure other microorganisms, their toxins or metabolites, either as part of a process verification for food suspected of presenting a hazard or as part of a risk analysis. Annex 2.5 of Regulation (EC) No 2073/2005 defines the microbiological criteria used for "vegetables, fruit and vegetable- and fruit-based products" (Table 4-6). This is limited to the measurement of *E. coli* colonies. Afssa referral no. 2007-SA-0174 also recommends measuring the presence of salmonella, aerobic micro-organisms at 30°C, and *Listeria monocytogenes*.

It is interesting to note that Sirsat *et al* (2013) showed that lettuces grown in aquaponics in greenhouses with a controlled environment had significantly lower concentrations of aerobic

Table 4-6. Regulatory thresholds for microbiological parameters in plant products.

Health parameter	Desired value	Unit	Regulatory source
Escherichia coli/g	< 100	CFU/g	Regulation (EC) no. 2073/2005
Salmonella/25 g	0	UFC/25 g	Afssa - Referral No. 2007-SA-0174
Aerobic micro-organisms at 30 °C/g	< 50 000 000	CFU/g	
Coagulase-positive staphylococci at 37°C	< 100	CFU/g	

flora, *E. coli* and faecal coliforms than lettuces grown conventionally in open fields. In addition, the same study showed that a simple preventive rinsing of the lettuces with a solution diluted to 2.5% acetic acid, after harvesting, was effective against most undesirable micro-organisms (*Salmonella* spp., *E. coli*). Microbiological analyses were carried out as part of the APIVA® project on various plants (edible part only) without any health risks being demonstrated.

The quality of the water, when and how it is used, and the characteristics of the crop all influence the potential of the water to contaminate the products. Fish are poikilothermic animals, which means that their body temperature varies according to changes in environmental temperature. In general, the body temperature of poikilothermic animals is too low to be considered optimal for the proliferation of most enteric bacteria that can affect human health (Fox *et al.*, 2012). This is because most human foodborne pathogens prefer relatively warm and stable temperatures. Zoonoses (infections of animal origin that can be transmitted to humans) originating from aquaculture products remain isolated cases (Hollyer *et al.*, 2009).

On the other hand, contamination of aquaculture ponds by warm-blooded animals remains possible, especially in an open environment: mice, birds (excrement containing *E. coli*), amphibians and reptiles (excrement containing *E. coli* and *Salmonella*). These animals must therefore be excluded from the production area. Humans can also be vectors of contamination if good hygiene practices are not followed. From the point of view of food safety with regard to plants grown in aquaponics, it is necessary to comply with "good agricultural practice".

Bacterial risks - mainly *E. coli* and *Salmonella* - can be easily avoided by good husbandry and cultivation practices (Hollyer *et al.*, 2009; Rakocy *et al.*, 2006; Fox *et al.*, 2012).

19.1.3. Regulations on nitrate levels in leafy vegetables

Regulation (EC) No 1881/2006 defines the maximum tolerated levels for certain contaminants in foodstuffs. Market garden plants are concerned by heavy metal levels, but also by nitrate levels in various leafy vegetables (lettuce and spinach). Table 4-7 summarises the standards to be met for lettuces and spinach, depending on the variety, the growing method (open field or under cover) and the growing season, where applicable.

studies show that soilless lettuces produced with fish farm effluent have nitrate levels below 2,400 mg NO_3^-/kg (Rico Garcia, 2009). Another publication shows that nitrate levels are the same in aquaponics and hydroponics, below 3,500 mg NO_3^-/kg (Pantanella *et al.*, 2010). These initial results provide encouraging information on the quality of products from aquaponics, but this parameter needs to be monitored in aquaponics given the lack of hindsight on this issue. The various analyses carried out on this parameter during the APIVA® project are encouraging, with over 99% of the samples analysed complying with the regulatory standards for this criterion.

19.2. Aquaponics and the organic label: two incompatible concepts ?

Aquaponics and hydroponics can be labelled "*Organic*" in the United States (the equivalent of

Table 4-7. Regulatory limits for nitrate levels in lettuce and spinach (Regulation [EC] No 1881/2006 setting maximum levels for certain contaminants in foodstuffs).

Foodstuffs concerned	Maximum levels (mg NO_3^-/kg)		
Fresh spinach (*Spinacia oleracea*)	Harvest from 1st October to 31 March		3000
	Harvest from 1st April to 30 September		2500
Preserved, frozen or deep-frozen spinach	/		2000
Fresh lettuces (*lactuca sativa L.*), grown under cover and in the open field, with the exception of "Iceberg" type lettuces.	Harvest from 1st October to 31 March	Lettuce under cover	4500
		Lettuce outdoors	4000
	Harvest from 1April to 30 September	Lettuce under cover	3500
		Lettuce outdoors	2500
"Iceberg" type lettuces	Lettuce under cover		2500
	Lettuce outdoors		2000

European organic labelling) since 25 January 2018 following an official clarification from the Department of Agriculture. This decision puts an end to a legal vacuum and to years of debate on the subject, with an almost philosophical confrontation between defenders of the link to the soil and supporters of soilless cultivation who consider this technique to be complementary to existing production tools.

In Europe, the situation is different. Article 4 of Council Regulation (EC) No 834/2007 of 28 June 2007 on organic production and labelling of organic products and repealing Regulation (EC) No 2092/91 states that "organic production shall use cultivation and animal production practices linked to the soil, or aquaculture practices respecting the principle of sustainable exploitation of fisheries".

Furthermore, according to Regulation (EC) No 710/2009 on "detailed rules for the implementation of Council Regulation (EC) No 834/2007 as regards organic production of aquaculture animals and seaweeds": "Recent technical developments have led to an increase in the use of closed recirculation systems in aquaculture; systems of this type depend on external inputs and are energy-intensive, but they make it possible to reduce waste discharges and prevent the risk of escape. In accordance with the principle that organic production should remain as close as possible to nature, the use of such systems should not be authorised for organic production, except, exceptionally, in the very specific case of the production phase in hatcheries and nurseries".

Regulation (EU) No 848/2018 of the European Parliament and of the Council of 30 May 2018 on organic production and labelling of organic products and repealing Council Regulation (EC) No 834/2007 specifies, in Article 3, the definition of an aquaculture facility: an aquaculture installation is "an installation, on land or on board a vessel, in which aquaculture takes place within a closed environment with a water recirculation system and dependent on a permanent input of external energy in order to stabilise the environment of the aquaculture animals" before laying down the rules applicable to organic production for these systems in Part III: "aquaculture animal production facilities with closed recirculation systems are prohibited, with the exception of hatcheries and nurseries

or facilities for the production of species used as feed for organic farmed animals". This regulation will apply from 1 January 2021, making it impossible for fish from recirculation systems to be labelled organic for several years.

Similarly, soilless plant cultivation is strictly prohibited under European organic regulations: the plant's roots cannot be in a solution or in inert material enriched with a nutrient solution under this label (no link to the soil). According to Écocert, "hydroponics contradicts the main principles set out in the European Organic Farming Regulation (RCE 889/08), which clearly states that the link to the soil is compulsory. Plants must be essentially nourished by the soil ecosystem".

Given that this notion of 'link to the soil' is a major intellectual positioning of the defenders of the European organic label, it seems difficult in the medium term to associate an organic label with aquaponics. So, failing that, we need to find another way of standing out and promoting aquaponics production, by putting forward arguments such as "local production", "short circuits", "circular economy", "sustainable agriculture", "no pesticide residues" or "zero pesticides", "zero antibiotics", and so on.

Faced with these regulatory challenges, aquaponics presents itself as a new model, or a combination of models, which requires a new framework to be thought through and built on the experience and projects already underway. The question of consumer perception of aquaponics is another important issue, as aquaponics also has to deal with societal challenges.

20. Societal challenges

20.1. The perception of soilless cultivation

20.1.1. The quality of soilless produce

Soil-less productivity often outstrips that of field crops. In economic terms, this sounds pretty good, especially as health is no more of an issue in soil-less than in open fields. But is the quality of the produce there?

Quality" refers to the nutritional characteristics of the plants produced - moisture content, sugar content, fibre, vitamins, minerals, carotenoids, etc. - depending on the type of plant and the expected quality. - depending on the type of plant and the qualities expected. A publication by Gruda (2008) summarises a large number of studies on this subject and suggests that soilless cultivation produces plants of similar quality to field crops in nutritional terms. The notion of quality also covers the organoleptic characteristics of the plants produced, and in particular their taste. The poor image of soilless growing is often due to the experience of most consumers: the lack of taste of supermarket tomatoes. In reality, growing practices are not necessarily responsible for this taste defect: it is perfectly possible to produce tasty soilless tomatoes just as it is possible to produce tasteless field tomatoes (Gruda, 2008). This taste defect in commercially grown tomatoes has various origins:

– Varietal selection has been carried out for decades on factors such as resistance to disease and pests, yield, earliness, texture, preservation and shock resistance, which has had an impact on the firmness and ripening speed of tomatoes. On the other hand, few selection efforts have been made in terms of product flavour;

– Tomatoes are usually harvested before they are ripe, sometimes when they are still green, whereas the flavours develop mainly in the final stages of ripening. Ripening takes place away from the plant and the fruit is no longer fed by the plant;

– the supply chain involves transporting and keeping tomatoes cold. This brutal cooling has an impact on the flavour of the product by inactivating the genes responsible for the production of volatile aromatic compounds (methylation phenomenon), which are mainly responsible for the flavour of these fruits (Zhang, 2016). Consumers exacerbate this effect by keeping tomatoes refrigerated.

In conclusion, the best way to eat a tasty tomato is to pick it when ripe and eat it quickly, provided that the variety is of any interest in terms of taste. The development of short distribution channels would seem to be a solution to the problem, as it would allow tomatoes to be picked when ripe and avoid storage and refrigeration prior to distribution as much as possible.

20.1.2. The quality of aquaponics products

Pantanella *et al.* (2010) have shown that aquaponics can modify the mineral composition of plants compared to hydroponics: lettuces grown in aquaponics are more concentrated in calcium, potassium, magnesium and sodium than in hydroponics, but also less concentrated in phosphorus.

Schmautz *et al.* (2016) showed that tomato plants grown in a decoupled aquaponics system (with supplementation of the nutrient solution with certain minerals) had yields and nutritional quality comparable to conventionally grown produce, although the levels of certain minerals (P, K, S, Ca, Mg, Fe, Cu and Zn) were above the minimum ranges 'required' for this type of plant (Schmautz *et al.*, 2016).

Few scientific publications refer to analyses of the nutritional and organoleptic quality of plants grown in 'pure' aquaponics, i.e. without the addition of supplementary nitrogen and phosphate fertilisers. There is still a great deal of data to be acquired on the quality of plants produced in 'unsupplemented' aquaponics, and one of the aims of the APIVA® project was to provide some answers to this question. Initial results from this project showed that vitamin A (beta-carotene) levels in basil grown in aquaponics were equivalent to those found in basil grown in hydroponics, in a study carried out in parallel on the same variety. A comparison with the values expected in 'conventional' basil from the ANSES Ciqual databases showed no significant differences. However, vitamin A levels in a sample of 5 lettuces were significantly lower in aquaponics than in the open field. The levels of calcium, phosphorus, magnesium and potassium found in basil or in lettuces grown in aquaponics and/or hydroponics were sometimes higher, sometimes equivalent, compared with conventional cultivation, while iron levels were generally much lower in aquaponics and hydroponics. It was also shown that the cultivation technique (NFT, *raft*, sub-irrigation table) often had a greater impact on product composition than the irrigation technique itself (hydroponics or aquaponics).

20.2. Consumer acceptance of aquaponics

Consumers are increasingly aware of the health benefits of locally and organically produced food. A certain proportion of the population have shifted their food purchasing decisions to these parameters, particularly in the most developed countries.

Consumer affect and certification criteria are crucial for the development of commercial-scale aquaponics (Miličić et al., 2017). It is therefore crucial to take into account their values, beliefs, the cultural and social norms in which they find themselves, the traditions to which they relate and, of course, their dietary tendencies. Aquaponics is a recurrent and fashionable topic in the media and social networks, but few studies have been done on consumer affect towards such products (Junge et al., 2017).

A recent study conducted in Berlin in 2017 on the acceptability of aquaponics in urban environments showed that only 28% of the consumer sample studied (386 people) approved of aquaponic fish and plant production in urban environments, and that only 27% of the same sample would be interested in purchasing this type of product if it were commercially available (Specht et al., 2016). Another study conducted in Malaysia (Tamin et al., 2015) found a high level of acceptability for aquaponics and the products made from it, and a high intention to buy. The same conclusions emerged in Romania, where the argument of the freshness of products was perceived as a good purchasing parameter (Zugravu et al., 2016). In Cuba, aquaponics is promoted by the government as an "eco-friendly" agricultural practice (Miličić et al., 2017).

A study conducted by Miličić et al. (2017) on a sample of 635 consumers in various European countries who responded to a survey (41% of whom were from Belgium) concluded that the perception of aquaponics was very good. Over 50% of respondents had never heard of aquaponics, while over 70% were already familiar with the hydroponics technique. The survey also showed that 17% of respondents were prepared to pay more for aquaponic products, with price increases of up to 40%. The study also found that 38% of respondents would choose fish from aquaponics over conventional systems, and that 23% would even be prepared to pay more for fish from aquaponics. Another interesting result of the study is that participants were more sensitive to arguments relating to local production, no pesticides/herbicides, no antibiotics: on average 54% of participants were prepared to pay more

for products meeting these criteria. This shows that aquaponics is not a sufficient argument in itself to differentiate products from what already exists on the market: it is necessary to back up the sales pitch with arguments such as 'health', 'proximity' or the 'circular economy' to succeed in creating value and getting people to accept price increases, while raising awareness of the practice of aquaponics could improve consumers' level of knowledge and thereby possibly redirect their intention to buy. It should be noted, however, that the level of purchase intention measured in consumer surveys does not always reflect the actual behaviour of these same consumers when making a purchase: these data should therefore be treated with a degree of caution.

20.3. The challenge of environmental integration

Aquaponics has developed under the impetus of a desire to move towards more sustainable aquaculture. This production concept opens up the prospect of new training courses, new professions, and even associations of players in the aquaculture and market gardening sectors within common structures to create a dynamic of short circuits in rural and urban areas. Aquaponics aims to address a range of issues affecting both society and the environment:

– improving the use of freshwater resources, at a time when this resource is scarce in some countries and, above all, very unevenly distributed around the world, while the world's population is growing at a sustained rate (+85 million per year);

– better use of inputs, compared with production systems that are often monospecific, where nitrogen and phosphate inputs are inefficiently assimilated by the organisms produced and threaten the integrity of soils and watercourses (eutrophication);

– preservation of natural resources and energy efficiency: the fertilisers released by the fish are a substitute for mineral fertilisers extracted or obtained by energy-intensive processes, while the state of the world's phosphorus reserves from mining is currently a source of concern for the future. They reduce the amount of fossil energy needed to manufacture the mineral nitrogen and phosphate fertilisers used in hydroponics. In addition, the volume of water required for aquaculture can potentially be an advantage in a greenhouse because of the thermal inertia of this body of water;

- the development of local commerce to limit transport and grey energy and minimise the distance between producer and consumer, while ensuring the origin and freshness of the products;
- optimising land use, as aquaponic greenhouses can be installed on non-arable land, on rooftops, but also on brownfield sites;
- food self-sufficiency in towns and cities, at a time when the expansion of urban areas and the world's population means that agricultural land cannot be extended;
- diversification of the fish farming and horticulture sectors, both of which are underdeveloped in France: for example, pre-existing above-ground horticultural greenhouses and fish farmers wishing to treat their waste more efficiently are potential candidates for the development of an aquaponics business, with lower initial investment than if they were starting from scratch;
- the development of more social links: through the development of shared urban systems, and by bringing together aquaculture and horticulture professionals, but also through the professional retraining and reintegration of people in difficulty.

Boxman *et al.* (2016) and Forchino *et al.* (2017) use a life cycle assessment (LCA) study to show that fish feed and high energy requirements are the main factors affecting the environmental balance of aquaponics.

20.3.1. Focus on the issue of fish feed

Fish feed is still largely dependent on marine resources (fish meal and fish oil) for its formulation. As fishing quotas are not expandable, we will have to move more and more towards the inclusion of alternative ingredients (based on plants, insects, algae, etc.). Transformed animal proteins (TAPs) - derived from the by-products of farmed mammals or birds - do not seem to be a promising avenue for the future in France, and are currently a dead end in terms of consumer image because of memories of the BSE or 'mad

cow crisis' in the 1990s, and the spread of sometimes unqualified statements by the media about the real reasons for this crisis.

Significant progress has been made since then. At present, to produce 1 kg of trout in France, a minimum of 1 to 3 kg (fresh weight) of 'forage' fish from industrial fishing is used, compared with 6 to 10 kg of prey fish consumed by these carnivorous fish in their natural environment (France Agrimer/CIPA). Research and new technologies are now enabling farmed fish feed to evolve towards greater use of raw materials of terrestrial origin, while maintaining the nutritional and organoleptic qualities of the fish. Feed manufacturers have already mastered the art of replacing 75% of fishmeal with a mixture of cereals, without altering the trout's growth, metabolism or immune system. INRA is currently selecting a line of trout capable of feeding on a 100% vegetable feed (oils and proteins).

Increasing use is also being made of processing co-products from fishing and/or aquaculture (in the region of 20% to 50% of fish oils and meals), which reduces the need for meals and oils from forage fish. Thanks to advances in nutrition and aquaculture selection, the target of 1 kg of trout produced for 1 kg of commercial feed consumed (conversion index) should be achieved by 2020, making sustainable aquaculture a reality (France Agrimer/CIPA). To achieve this, it will also be necessary to find solutions for replacing fish oils: vegetable oils give good results, but the lipid spectrum of fish is affected, with a significant drop in the composition of omega 3 polyunsaturated fatty acids EPA (eicosapentaenoic acid) and DHA (docosahexaenoic acid), which are particularly sought after in human nutrition. Here again, research has shown that it is possible to restore normal lipid spectra with a fish oil-based finishing diet at the end of the production cycle. New technologies have also made it possible to produce omega-3 polyunsaturated fatty acids using transgenic camelina (Betancor *et al.*, 2015), an oilseed plant not intended for human consumption. The fact that these 'transgenic' fatty acids end up indirectly on consumers' plates raises the question of the acceptability of this practice, and limits this avenue for progress.

Other avenues of research are being seriously considered and have already been tested with very encouraging results, such as insect meal (authorised for use in aquaculture feed in Europe since 01/07/2017), bacteria and yeast, and micro- and macro-algae.

The problem mentioned above, linked to the feeding of farmed fish, raises other questions, linked to European dietary traditions. Our current food model is based largely on animal proteins (meat, milk, fish). Fish consumption in Europe is tending to increase, but not for just any species: mainly salmon, trout, sea bass, sea bream, turbot, etc. What these fish species have in common is their naturally carnivorous diet, and therefore their appetite for animal proteins. In Asia, on the other hand, the tradition is to eat fish with a predominantly plant-based or omnivorous diet (carp, tilapia, catfish, etc.). These fish - with their low trophic level - are also very popular in the United States.

20.3.2. Focus on the problem of dependence on carbon-based energies

The power consumption of an aquaponics system is another major issue affecting the environmental impact of aquaponics, which is, after all, a highly technology-dependent farming practice. A large-scale aquaponics system necessarily requires pumping systems to circulate the water, oxygenation and mechanical filtration systems to ensure impeccable water quality for the fish and plants, and in some cases greenhouse heating systems, or even lighting, for winter operation in temperate climates. The energy requirements inherent in the operation of the system depend not only on its configuration (design, species selected, technologies used) but also on its geographical location and the associated climate.

The adoption of renewable energy technologies, ecological heating solutions (wood-fired boilers or agropellets, geothermal heat pumps, recovery of industrial thermal waste in agrothermal parks, biogas production, etc.) and efficient and durable lighting (Led neon lights now have a lifespan of over 100,000 hours) are essential solutions if aquaponics is to have any legitimacy in environmental terms (Junge et al., 2017), the only obstacle today being the price to pay for these alternative solutions, 2017), the only obstacle today being the price to be paid for these alternative solutions... It may also be possible to replace pumps with *airlift* systems for small-scale fish farming systems (Barrut, 2011).

21. Economic challenges

Regulatory, social and environmental issues are major concerns, but the economic challenge is probably the most crucial. The economics of commercial aquaponics systems in temperate climates have not been extensively researched to date. Profitability and yield per square metre of soil surface are the best indicators for comparing the economic efficiency of different growing systems (hydroponics, aquaponics, open land; aquaponic system A, aquaponic system B, etc.). This profitability per square metre is influenced by:

- markets (product added value, supply/demand correlation). According to Xie *et al* (2015), the ideal strategy is to supply local restaurants or small local markets directly. Large-scale distribution is not *a priori* a source of profit for this growing method under the current conditions and technical requirements;
- local climatic parameters (seasonal availability of daylight, temperatures);
- the cost per square metre, itself influenced by the values of investment, operating costs (energy, inputs, staff) and maintenance costs. Based on initial estimates and the various studies, the expenses that weigh most heavily on operating costs appear to be staff, energy and fish feed;
- the added value of farmed fish species and cultivated plants.

21.1. The paucity of technical and economic studies in the literature

We are only just beginning to master aquaponics from a technical point of view, and now we need to compare these production systems with economic analyses. There is little scientific literature on the economic feasibility of aquaponics, and for good reason: the practice is still in its infancy.

The first technical and economic study was carried out as part of the University of the Virgin Islands' experimental programme at the St Croix

Experimental Station (Bailey *et al.*, 1997; Rakocy *et al.*, 2006). The results show that the tropical climate is particularly well suited to plant cultivation (light, temperature), that the fish used are particularly hardy (tilapia), and that, in this region, the plants produced (lettuce, tomatoes, basil) have significant added value (Engle, 2015). Bailey *et al.* (1997) analysed data from three sizes of farm producing tilapia and lettuce and showed that economies of scale could be significant, and that farms that were too small did not provide sufficient returns on investment. In a similar climatic context, Tokunaga *et al.* (2015) carried out a study on three farms located in Hawaii, which respectively comprised: 1,070 m² of crops for 68 m³ of livestock, 1,141 m² of crops for 27 m³ of livestock, and 2,657 m² of crops for 340 m³ of livestock. All three of these small-scale farms were profitable, and the technical and economic simulation carried out by the authors on the basis of these 'model' farms suggests that aquaponics is a viable option for producing vegetables and fish for local markets. The authors indicate that the main constraints on aquaponics in the Hawaiian context are high initial investment costs, labour intensity (18% of production costs), dependence on electricity (28% of production costs), the cost of fish feed (11% of production costs) and the very significant impact of product prices on economic results. On average, energy and fish feed represent 23% and 11% of production costs respectively on these farms, while labour represents only 18%. Other risks to system profitability include plant losses due to pests and diseases. Baker (2010) also analysed an integrated tilapia and lettuce production system in Hawaii and found it to be technically feasible and profitable. The author concludes that there is no economic interest in integrated lettuce production with tilapia compared with conventional lettuce monoculture, while noting the ecological benefits of aquaponics. All these studies have the tropical climate in common and cannot necessarily be generalised to temperate climates.

In an international aquaponics study of 257 aquaponic farms, Love *et al.* (2015) found that tilapia was the most common fish species raised in aquaponic systems, while vegetables and herbs (basil, lettuce, tomatoes, leafy greens, kale, chard, bok choy, peppers and cucumbers) were the most commonly grown plants. The study showed that many aquaponics farms were not financially viable. Of the 257 farms studied (208 of which were located in the United States), only 31% reported that their business was profitable during the last 12 months of the study. Chaves *et al.* (2008) studied the integration of hydroponic tomatoes into a recirculating system producing catfish in Scotland. A rate of return (ratio between the income obtained and the initial capital outlay) of 27.32% was achieved, a financial result fairly close to what can be obtained for the production of catfish alone. The authors point out that partial purification of fish farm effluent would provide an additional benefit in terms of social cost. Lapere (2010) carried out another techno-economic study in South Africa, and concluded that the high capital and operating costs made it difficult to achieve profitability unless large-scale structures and the resulting economies of scale were targeted. Quagrainie *et al.* (2017) conducted an economic study to compare the economic efficiency of hydroponics and aquaponics in a cold temperate climate in the Midwest region of the United States. The authors indicate that aquaponics requires higher investment and production costs than hydroponics, with a lower plant yield. The condition for aquaponics to be profitable is, according to their study, linked to the labelling of products from this system, with a selling price around 20% higher than that of conventional products. In this model, fish feed, energy and labour accounted for 3%, 21% and 49% of production costs respectively.

Although only a few studies have been carried out in continental and Mediterranean regions, the existing literature on the economic aspects of aquaponics largely covers tropical areas and the fish and vegetable species mainly adapted to these climates. Consequently, further research is needed in temperate contexts, particularly in colder climates.

21.2. Technical and economic studies of three low-tech aquaponic systems

Heidemann (2015a) carried out an analysis of three commercial aquaponic farms in the United States that agreed to provide economic data: James Rakocy's UVI system, Lily Pad Farm in Texas and Traders Hill Farm in Florida. While the results cannot necessarily be transposed point by point to the European situation, they do provide some key indicators. The following is a summary of this very detailed study.

21.2.1. The University of the Virgin Islands system - UVI

James Rakocy carried out research on aquaponic systems between the early 1980s and 2010 at the University of the Virgin Islands. His work led to the development of design data based on a ratio of fish feed to plant area. The UVI experimental pilot design has been widely adopted by other commercial projects.

21.2.1.1. TECHNICAL ASPECTS (HEIDEMANN, 2015A). The UVI experimental system operates in the open air and is located at the University of the Virgin Islands. It comprises a tilapia farm in a loop with vegetable cultivation on *rafts*. The aquaculture components of the UVI system are covered by bird netting and a tin roof. The system occupies a total surface area of 500 m². The system consists of 4 rearing basins (31 m³ in total) and 6 plant culture supports (220 m² in total surface area). The recirculation flow rate is 22 m³/h, ensuring just under one renewal per hour in the fish rearing tanks, which is often sufficient for robust fish such as tilapia. The rearing density reaches 60 kg/m³. Daily feed intake averages 12 kg per day. The water leaving the fish ponds is enriched with potassium (K), calcium (Ca) and iron (Fe) to avoid deficiencies in the plants. Potassium and calcium are supplied in the form of hydroxides, which also serve to raise the pH to compensate for bacterial acidification in the biological filter.

21.2.1.2. ECONOMIC ASPECTS. Various varieties of plants are grown there (basil, lettuce, okra, melon, etc.) with a commercial objective. The UVI system produces 5 tonnes of tilapia a year (580 kg every 6 weeks, i.e. 160 kg of tilapia/m³ of water/year, with rearing densities of around 60 kg/m³), as well as 1,400 crates of lettuce (25 lettuces per crate, i.e. 35,000 lettuces a year), 3 tonnes of basil and 2.9 tonnes of okra a year. The products are marketed on an island where there is little competition in the market-gardening sector, and where the limited supply means that products can be sold at a high price. This niche market situation makes it easier for

the farm to make a profit, with selling prices of €5/kg for tilapia, €20/kg euros per kilo for basil and €0.75/piece for lettuce (Rakocy *et al.*, 2004), prices well above the minimum profitability estimated by Rakocy for this aquaponics system. On average, 81% of the income generated by the UVI system comes from the sale of plant products and only 19% from fish.

The investment required to build the pilot amounts to €33,000. The experimental UVI system would generate a net profit of 39,000 euros from the second year (Heidemann, 2015a), a result obtained without deducting taxes.

21.2.1.3. SPECIAL FEATURES. It should be noted that the special geographical position of the Virgin Islands gives rise to specific features that need to be taken into account:

– the climate is tropical, with temperatures above 20°C all year round, which favours good plant growth evenly throughout the year. What's more, this climate makes it possible to set up an outdoor system, which is inevitably less expensive than under glass (Rakocy *et al.*, 2006);
– the island status of the Virgin Islands makes the market and therefore selling prices very specific. An aquaponics system that enables fresh vegetables to be produced locally all year round is easily adapted to a market where most products are imported from the rest of the American continent (Rakocy *et al.*, 2004);
– the cost of importing certain materials and inputs (such as fish feed) is high, given the island nature of the area.

21.2.2. Lily Pad Farm - LPF

Adam and Susan Harwood are the owners of Lily Pad Farm (LPF) in San Marcos, Texas. They have been researching the field for some time and have spent years planning the construction of their farm. Adam has also worked in aquaculture, which is a real advantage given the technical nature of this sector. The fish farming part of the business is very important and is not simply a supplier of fertiliser: profitability is sought in this area.

21.2.2.1. TECHNICAL ASPECTS (HEIDEMANN, 2015C). The farm is surrounded by ranches and covers an area of around 8,000 m². The aquaponics section consists of three plastic tunnel greenhouses (30 m long by 9 m wide, i.e. 270 m²), each containing a production system modelled on the UVI system.) Each system consists of two 27 m³ fish rearing tanks (54 m³ in total), a conical-bottomed settling tank, two filtration tanks, a degassing column and two hydroponic tanks, each measuring 50 m². The rearing tanks each hold around 1,450 tilapia, with rearing densities reaching 70 to 80 kg/m³. Each of these hydroponic beds contains 18 floating *rafts*, each measuring 2.8 m² and each drilled with 48 holes (i.e. around 17 holes per m²). These systems are also built inside greenhouses, which require additional heating for several months of the year. However, this system does not contain a biological filtration system, which represents a relatively significant saving in terms of investment and production costs associated with water mixing: producers rely solely on the plant component to purify the ammoniacal nitrogen released by the fish; this is possible with tilapia, which is highly tolerant of physico-chemical conditions, but much more risky with fish such as trout, perch, pike-perch or sturgeon.

21.2.2.2. ECONOMIC ASPECTS. The LPF system is capable of producing over 1,400 fish per month and an average of 1,500 plants of herbs or lettuce per week, all year round. For around four to twelve weeks of the year, weather conditions affect yields in this geographical area (too cool or too hot). Each tilapia is sold for €4.8 a piece ('portion' fish), while each plant is sold for €2.8 a piece on average (the plants grown are not specified): these prices are high overall and help the farm's bottom line.

On average, 71% of the income generated by the LPF system each year comes from the sale of plant products and the remaining 29% from fish sales.

The investment for the three greenhouses was around 27,000 euros, or 9,000 euros per 270 m² greenhouse. Each complete aquaponics system in the greenhouses cost around €32,300. For each unit comprising around 18 m³ of rearing and 100 m² of plant cultivation, an investment of 41,300 euros was required, giving a total of 123,900 euros. In the first year of production with the three aquaponics systems in place, LPF generated sales of €127,516, rising to €200,000 in the second year. From the third year onwards, the net result of the business was 100,000 euros/year, after deducting all variable and fixed costs, but excluding taxes.

21.2.2.3. SPECIAL FEATURES. The LPF managers are adamant that commercial aquaponic farms should be built, not bought: their DIY skills have resulted in substantial savings. Their advice is to start with a small farm and expand as experience is gained. This farm is run on a family scale, and living at the place of work saves a lot of money.

21.2.3. Traders Hill Farms - THF

Traders Hill Farms (THF) is part of a 40-hectare farm owned by the Blaudows family in Florida. The Blaudows decided to set up a sustainable business that could foster economic development in their community and create local employment. THF was launched in December 2012 and its 'Centre for Sustainable Agricultural Excellence and Conservation' was launched in the same breath in November 2013. The centre has been developed to become an education and awareness institute for individuals interested in learning more about sustainable farming techniques. Through the centre, they work with local farmers to help them plan, build and operate their own aquaponics systems, providing assistance throughout the process.

21.2.3.1. TECHNICAL ASPECTS (HEIDEMANN, 2015D). The owners of THF meticulously studied and experimented with different aquaponics systems until they came up with one that was functional, adaptable and easily reproducible. Again, this system is inspired by the UVI system, but it has many differences. THF's aquaponics system is housed inside an old henhouse with numerous transparent roof panels, allowing sunlight to reach the plants. In winter, artificial lighting is used sparingly. The henhouse also houses a tiled wood-burning stove, used for heating during the coldest months of the year. The THF system consists of 4 rearing tanks, each measuring 5 m³, representing a fish production volume of around 20 m³, 2 settling tanks, 2 polypropylene foam filter mats (Matala filters) to supplement mechanical filtration for the finest particles, 2 fluidised bed biological filters filled with filtration media, a degassing tank, 3 hydroponic tanks measuring 30 m long by 1.2 m wide, giving a usable surface area of 108 m². The water flows from the rearing tanks to the settling tanks: one settling tank for every two rearing tanks. The water then passes by gravity through the Matala filters, then through the biological filters. From here, the water accumulates in a tank dedicated to degassing before being fed into the hydroponic culture tanks. A pump then pumps the water from the plants to the rearing tanks. A blower is used to aerate the entire system.

21.2.3.2. ECONOMIC ASPECTS. The Traders Hill Farms system is capable of producing up to 25,000 plants and 1,000 fish per month. During the few weeks of summer, it can be too hot in Hilliard, Florida, for some plants to achieve optimum yields. Traders Hill Farms sells its production at an average of €0.8 to €1/plant and €3.9/kg of tilapia.

Around 91% of the revenue generated by the THF system comes from the sale of plant products and just under 9% from fish sales.

A total of 84,000 euros was invested in the construction of this aquaponics system, with 81% of the costs attributed to the rearing tanks and culture vats. It should be noted that the building was already in place, so these costs do not include the construction of a greenhouse. In the first year, the company generated sales of €110,000; in the second year, €137,000; and in the third year, €245,000. From the third year onwards, the company's net profit was €109,000/year, with tax included in the calculation each time.

21.2.3.3. SPECIAL FEATURES. It is important to note that the managers of Traders Hill Farm did not have to buy the building housing the aquaponics system, as they already owned the land, which represented a significant saving. In addition, heating costs are minimal for this business located in North Florida. All these benefits have to be factored into the bottom line. Since the economic study carried out by Heidemann in 2015, the structure has grown to include several greenhouses with a production area of 5,000 m².

21.2.4. Conclusion on case studies

The clear trend that emerges from these case studies is that the plant compartment often produces the largest share of the income earned by aquaponics producers. Despite their limited size,

the systems studied were profitable at the time of the study, when the scale of plant production was between 100 and 300 m² for the three cases studied. Note the absence of a greenhouse for the UVI open-air system, the presence of an unenclosed, inexpensive plastic greenhouse for LPF, and the use of an existing building for THF. The managers of the various companies already owned the land, which meant that land costs were not taken into account in the economic study.

Most of the start-up costs associated with these systems can be attributed to the equipment needed to build them, such as fish tanks, hydroponic tubs, air blowers and pumps. These initial costs can significantly affect the bottom line, particularly in the first few years of production. Self-build can be very cost-effective.

The income generated by the sale of products alone is enough to provide a decent income for these producers and their families. Each system also generates enough profit to reinvest the funds in the business. The return time (or payback period, i.e. the time needed for the projected cash flows generated by an investment to recoup the initial investment cost) was estimated by Heidemann (2015a) at 22 months for UVI, 41 months for LPF, and 31 months for THF.

Feedback is a valuable source of information, but it should also be treated with a great deal of hindsight. Producers replicating one of these systems should expect that the economic results can vary considerably depending on the context: the success of an aquaponics business depends on the technical knowledge and experience of the operator, the geographical and climatic environment, the economic environment and the ability to stand out from the crowd.

The fact that you already own a plot of land, or even a production greenhouse that is already operational or just needs a few renovations, is a major advantage.

21.3. Presentation of "high-tech" aquaponic systems viable

The examples detailed below are intended to present a few "remarkable" aquaponic farms that are currently economically viable; this list is not intended to be exhaustive. In most cases, these farms have required substantial fund-raising, made possible by investors who are convinced by the concept of aquaponics. Unlike the *low-tech* structures presented in the previous paragraph, no in-depth study has been made public on the operation of companies of this type, as economic data is often highly confidential. The data presented in this book therefore comes from the websites of these companies, or from press articles.

21.3.1. Cultures Aquaponiques M.L. inc - Canada - Sainte-Agathe-des-Monts

Marc Laberge was one of the pioneers of commercial aquaponics in Canada when, in 2004, he began growing Boston lettuce in *rafts* above a rainbow trout pond in Sainte-Agathe-des-Monts in the Laurentians, Quebec. "Since its first sales in June 2005, the company has produced more than 3.8 million heads of all-natural lettuce and has managed to carve out a place for itself in the Boston lettuce market in Quebec, despite the fact that the world's largest producer of hydroponic lettuce using traditional chemical methods is just half an hour's drive away," states the company's website.[8]

Marc Laberge says that it's not easy to set up a viable aquaponics business: among the obstacles he cites are ruthless markets and the difficulty of getting products recognised as organic. Faced with these difficulties, in June 2014, after 9 years in business, Marc Laberge decided to sell his business and devote himself to consulting. His farm was eventually bought by an entrepreneur who decided to go into partnership with him and share production costs. In 2015, this commercial system produced 48 tonnes of lettuce (5,000 to 6,000 pieces per week) and 7.5 tonnes of trout per year. Plans to expand the business are under way.

21.3.2. Urban Organics - United States - Saint Paul

Back in 2014, entrepreneurs Kristen and Dave Haider, Chris Ames and Fred Haberman decided to buy and renovate an old derelict building, a former brewery called "Hamm's Brewery" in Saint Paul, Minnesota (for $150,000), their goal being to install a commercial aquaponics system. They realised their dream and founded Urban Organics. Their first system was around 800 m² and produced aromatic herbs (mainly mint, basil, watercress and lettuce) and tilapia. Urban Organics joined forces with Pentair, a company specialising in water management and filtration, to accelerate its growth. Together, the two partners are building an 8,000 m² farm capable of producing 124 tonnes of fish and 200 tonnes of plants a year, all with organic certification, since organic wholesalers are their main customers. They grow a variety of herbs and leafy greens using LED-lit *rafting* on three superimposed levels, and also raise salmon and Arctic char to take advantage of Minnesota's cold climate. Rearing cold-water fish means lower energy consumption than in their first pilot system, where they chose tilapia, a robust species that requires warm water, between 25 and 28°C. Mastering salmonid farming required a high level of technical expertise (particularly in terms of filtration), which Pentair was able to provide as part of their collaboration. The future strategy is to develop their model in other regions of the country.

21.3.3. ECF Farmsystems - Germany - Berlin

In 2012, Christian Echternacht and his partners at ECF Farmsystems developed a small-scale farm concept consisting of a greenhouse on the roof of an industrial container and a fish farming system inside the container. This system was not intended to be profitable, and served mainly as a laboratory. "This technique can only be profitable over a large surface area" (*says* Christian Echternacht). That's why a large-scale farm was completed in Berlin in 2014, covering an area of 1,800 m² and costing €1.2 million. The vegetables grow in a large greenhouse, right next to the fish tanks. "We heat the greenhouse very little, no more than 5°C in winter, and we don't use artificial light," adds Christian Echternacht. Each year, we plan to produce 30 tonnes of fish and 35 tonnes of vegetables. Berliners can buy the produce in the shop next to the farm, or have it delivered by subscription. "We distribute through several sales channels. On the one hand via our own market stall in Markthalle 9 in Berlin Kreutzberg, and on the other through various retailers. We also supply various restaurants. Thanks to this, we have a good diversity of customers". Once the concept has proved its viability, the aim, like Urban Farmers, is to market it and build several farms across Europe.

21.3.4. GrowUp Urban Farms - England - London

The company started out in the same way as ECF Farmsystems, with a container farm system called the GrowUp Box. Since 2015, founders Kate Hoffman and Tom Webster have been developing an urban commercial farm project by creating Unit 84, a multi-storey aquaponic farm under artificial lighting on a brownfield site in a suburban area. The farm has 560 m^2 of production space (thanks to the optimisation of space offered by vertical cultivation) and can produce 20 tonnes of plants and 4 tonnes of fish a year. To date, the products are sold mainly to the local catering market.

Unit 84 is just a 'trial run' for its founders, serving primarily as a pilot unit for the future development of an even larger-scale farm; the economic aim of GrowUp Urban Farms is to supply large supply chains, targeting the '*baby leaves*' and salad markets. In their view, "agriculture is happening on such a huge scale that [if they want] to have an impact, [they need] to do something that fits into the supply chains that already exist". They are not looking for a 'high-end', niche or 'organic' market, but simply to produce close to major food centres, such as supermarket distribution centres and wholesalers, so that they can deliver large quantities of vegetables quickly and minimise the time, cost and environmental impact of transporting these perishable products. According to the co-founder, the aim is to "provide the salad and fish fillets on supermarket shelves: the proteins and salads that people eat every day".

21.3.5. Superior Fresh - United States - Hixton

Superior Fresh is the world's largest aquaponics facility, located in rural Hixton, Wisconsin. The giant 15,000sq m aquaponics farm (including 11,000sq m of plant cultivation) opened in August 2017, with the aim of "changing the world" through sustainable farming and healthy eating. Production is said to be 900 tonnes of plants and 80 tonnes of salmon a year. According to president Brandon Gottsacker, plant and fish products supply retailers, restaurants, schools and hospitals in the Midwest. Superior Fresh plans to further expand its distribution of leafy greens and began selling Atlantic salmon in early 2019.

21.3.6. Bigh Farms Slaughterhouse Farm - Belgium - Brussels

The Abattoir farm is a spectacular example of an aquaponic installation on the roof of a building in a peri-urban area. It was launched in Brussels in 2018. It was developed in partnership with ECF Farmsystems, with the aim of creating a commercial model for integrated aquaponics on the roof of an urban building. It was designed by Steven Beckers, architect and founder of the consultancy firm Lateral Thinking Factory, which has been supporting the implementation of the circular economy in the property sector for many years. The structure comprises 2,000 m^2 of market garden greenhouses linked to a fish farm, and 2,000 m^2 of outdoor vegetable garden, all at an investment cost of around €2.5 million. Forecast annual production is 40 tonnes of plants (aromatic herbs, micro-sprouts, vegetables and fruit) and 35 tonnes of striped bass. A network of partners and professionals has been set up around the start-up, with the idea of distributing the produce via short distribution channels, capitalizing on the current trend among city dwellers to 'consume locally'. Eventually, Bigh Farms aims to create a network of similar farms in the heart of Europe's major cities. "In the long term, the city becomes a solution if we seek to have a positive impact at every level - energy, water, air quality, biodiversity, material resources - while creating jobs and integration," says Steven Beckers.

21.4. The importance of context

The potential advantages of commercial aquaponics systems are numerous: high yields, shorter growth cycles than open-field cultivation, a degree of uniformity in product quality, good traceability, an extension of the production period (with greenhouse production), conservation of resources and a reduction in inputs. But that's not enough. The key is to develop viable economic models around aquaponics and to master all aspects of production, from the technical side to the sales strategies that will be specific to each producer in a given geographical, climatic and competitive context.

As we have pointed out, the current literature is not sufficient to provide a critical assessment of the economic efficiency of aquaponics.

The practice is still at an early stage of development, and the technical aspects alone are only just beginning to be mastered. In addition, it is extremely complicated to compare the efficiency of two given systems, given that there is no standardisation of protocols for measuring the various parameters measured, and that the context will be totally different from one region to another. For example, energy costs will be lower in a tropical country, whereas they will be a major cost in temperate regions in winter. Much of the literature is based mainly on hypothetical situations and specific frameworks. In the absence of realistic data on existing farms and in a European context, such projections are often over-optimistic because they simply do not take account of unforeseen circumstances.

A system designed in European latitudes will almost certainly have to operate in a greenhouse or inside another building or structure. This would allow a farmer to use auxiliary heating during the coldest months of the year, and to use climate control throughout the production year if necessary.

21.5. Key success factors

Although setting up an aquaponics business still involves certain risks inherent in the complexity of the *process* and the costs involved, whether in terms of investment or production costs, the experience of various "aquaponiculturists" has revealed a number of key success factors that are essential to follow.

21.5.1. Mastering the technical aspects

Aquaponics is not an easy business, where all you have to do is feed fish to grow plants. The beginner should be aware of this. It seems obvious, but if you're planning a commercial aquaponics production project, it's essential to master the technical aspects and understand all the issues involved: microbiology and the nitrogen cycle, fish zootechnics, horticultural skills, the operation of a hydraulic system, sanitary aspects, safety issues relating to electrical circuits, regulatory constraints on plant and fish species, etc. Finally, aquaponics is a complex ecosystem, which means that you need to know how to

maintain a certain balance in the physico-chemical parameters of the water, with acceptable levels for the plants, fish and bacteria. Without measured and scientifically reasoned sizing, the risks of failure are real.

In its routine operation, an aquaponics system will need to be managed by individuals with dual skills: aquaculture and soilless horticulture. Each of these fields is a world of knowledge to be acquired.

21.5.2. Building a viable business model

The examples of aquaponic farms given above show that it is now possible - but not easy - to create an economically viable structure based solely on production. The first step is to draw up a forward-looking production plan, underestimating the achievable turnover as much as possible and making pessimistic assumptions about the system's investment, operating and maintenance costs. Remember Murphy's Law: "anything that can go wrong, will!

Plant and fish production also needs to be adapted to the climate and, above all, to local market demand (market research is essential). It is sometimes necessary to resort to heating if plants are produced out of season, or if 'exotic' fish are reared. It is preferable to focus on aquatic plants and animals with high added value, and to ensure rapid rotation of short-cycle plants. It is possible to mix different hydroponic techniques (*rafts*, substrate, vertical/horizontal NFT, inert substrates) to make the most of the variety of products grown. It can be cost-effective not to germinate your own seeds, but to find a local supplier of young plants to save time on production cycles. The same applies to the acquisition of fry, which must be of excellent quality to avoid any health problems with fish production.

Most aquaponics growers cultivate plant species with a high economic value, such as rare vegetables, aromatic and medicinal plants and "young shoots". The choice of fish species depends more on a compromise between regulations, the availability of fry, the characteristics of the renewal water, local demand and the expected selling price. Producers can increase the added value of their products by processing them (filleting or even smoking fish, processing basil into pesto, for example). This requires investment in specific tools and additional jobs, or subcontracting to a food processing plant as part of a local approach.

Lastly, it would seem wise to minimise initial investment by rehabilitating derelict buildings/industrial wastelands/abandoned farmland.

Several strategies are therefore possible, depending on the product on which you want to maximise sales:

–   strategy 1: the fish compartment in itself can be a simple alternative nutrient source to fertiliser for the majority production (the plants) without being profitable on its own. Most commercial aquaponics farms earn most of their income from the sale of plants (Love *et al.*, 2015). This strategy seems rather appropriate for a market gardener wishing to limit the use of chemical fertilisers and at the same time improve his image with consumers. In practice, however, this strategy is rarely viable, because of the investment required to design the fish farm, unless the focus is on producing plants with very high added value (medicinal or cosmetic plants, rare vegetable plants for top chefs, etc.) or on a very large scale;

–   strategy 2: on the other hand, the plants can be considered as simple water purification tools in recirculated aquaculture circuits with a commercial objective (with decoupling of the two compartments), resulting in additional income for the fish farmer. In this case, the aim is to make use of livestock effluent, without optimising the sizing for the plant compartment. This strategy would appear to be best suited to an aquaculture operator wishing to discharge effluent with a lower nitrogen and phosphorus content, in order to comply with effluent regulations;

–   strategy 3: you can opt for a fully integrated production strategy, with the aim of achieving double profitability: fish farming and market gardening compartments. You can choose a "coupled" or "decoupled" aquaponics system, the latter allowing you to minimise risk by ensuring that each compartment can operate independently in the event of a problem. This strategy is the safest economically speaking, but also the most complex to set up because of the scale of production required to make the fish

farming part profitable on its own. The advantage is that it is possible to use a combination of two professionals, each with their own expertise. In this way, aquaculture and market gardening professionals can complement each other in a win-win relationship, where the fish farmer entrusts the management of his effluents to a market gardener and the market gardener manages the irrigation of his above-ground crops with a natural fertilising solution, making significant savings on fertiliser. The water treatment systems (mechanical and biological) can be co-financed by the two companies to share the investment costs. Thanks to this kind of collaboration, there is no need to merge the economic models of the fish farming and market gardening compartments. Each production workshop is independent, and must therefore be profitable on its own: the aim is simply to lower production costs for both companies.

21.5.3. Stand out

We need to be aware of the difficulty of competing on price with products from aquaponic systems compared with those from intensive conventional techniques. It's a fact that producing plants in open fields costs less than producing them in aquaponics, even though aquaponics offers a higher yield most of the time. This means that the selling price of fish and plants must be higher than in conventional farming, or at least that the producer must ensure that there are as few intermediaries as possible between the farm and the consumer. We need to come up with an effective marketing argument for communicating about aquaponics products to raise their profile and explain how they differ from conventional products.

It is possible to stand out mainly on the basis of image: preserving water resources, reducing nitrogen and phosphate pollution, 'natural' plant production, reducing the carbon footprint by adopting short distribution channels for products, creating 'green' jobs, the absence of chemical fertilisers and pesticides in the plant production process, the absence of antibiotics in fish production, the 'local' nature of products, etc.

It is also possible to communicate the 'reasoned' aspect of aquaponics, a vision that seeks to find a compromise between the productivity objectives of conventional farming and respect for the environment, which tends to bring it closer to organic farming without adopting all the constraints and therefore without being able to claim the label. In this sense, aquaponics is perfectly rational, free from chemical nitrogen and phosphate fertilisers, and part of a sustainable development approach.

Quagrainie *et al.* (2017) claim that aquaponics can become more profitable than hydroponics, provided that the products can be labelled organic and sold at 20% higher prices than conventional products. As we have seen, organic is not an option in Europe for products from this production system, but certain regional labels promoting 'short circuits' do exist and can be used to add value to products and to focus communication on a local and healthy production approach, with no transport and no refrigeration of products before sale. Another example is the "zero pesticide residue" label, which is becoming increasingly popular. Similarly, there seems to be room for the official creation of a label specifically for aquaponics and, more generally, for 'integrated' agriculture.

21.5.4. Developing a short-distance niche market

Many consumers are increasingly interested in buying locally grown food, and want to know how their food is produced. They also increasingly want to buy directly from a farmer with whom they have established a relationship of trust. These consumer profiles are often prepared to pay more for these attributes (Heidemann, 2015a): they are therefore customers to target if we hope to achieve a viable economic model for aquaponics. It is also relevant to look for niche markets (gourmet restaurants, hotels, school or company canteens, etc.) given the difficulty of making aquaponics products compete with conventional products in terms of selling price. Direct sales to consumers in markets or directly on the farm in agritourism and group purchasing systems (AMAP) are the preferred distribution channels. Some producers in the United States, such as Greens and Gills, have even gone as far as online sales and home delivery.

Heidemann (2015a) showed, based on a sample of 39 commercial producers in the United States, that around 31% of production was sold directly at the farm, 18% to restaurants and 16% at markets. This is in line with the preference for these short circuits to avoid transport costs and get closer to the consumer, a condition that seems almost mandatory to hope for a profitable aquaponics business.

21.5.5. Don't neglect ancillary activities

The development of ancillary activities as sources of additional income is very often the salvation of many existing aquaponic businesses (Heidemann, 2015a). For some, these activities are even placed at the heart of the project: the business becomes a place for vocational training or dedicated to education, with the farm opening up to a wide public (agritourism). Others are looking to develop ancillary services, such as the supply of small-scale turnkey aquaponics kits for *high-tech* garden enthusiasts, as well as the supply of equipment and even design and consultancy services. Heidemann (2015a) showed on a sample of 39 commercial producers in the United States that on average 39% of the total annual household income of these producers could be attributed solely to their commercial aquaponics products and services. More than 23% of these growers reported that their aquaponics products and services accounted for more than 75% of their annual income. Nearly 31% indicated that their aquaponics products and services accounted for more than 50% of their annual household income.

22. Conclusion

A number of commercial projects have emerged in France in recent years, thanks to adventurous entrepreneurs from a wide variety of backgrounds who want to be pioneers in aquaponics. However, the legal framework for the emergence and development of aquaponics is still in its infancy and vague, and it would seem useful for regulations to fill this legal void. One thing is certain, however: there is currently nothing to prohibit the sale of soil-less cultured plants as long as they comply with regulations governing the sanitary and microbiological quality of products, but there is also nothing to indicate that this is authorised. From a purely regulatory point of view, irrigating plants with fish farm effluent is truly unprecedented, and this water from fish farms does not fall into any category of water authorised for irrigation in market gardening, which could one day raise questions; questions that need to be anticipated by confronting the health regulations inherent in conventional market gardening products, as well as those governing the quality standards for water used for irrigation.

More generally, the development of agriculture and livestock farming (including aquaponics) in an urban environment raises regulatory issues relating to the management of organic waste (livestock sludge) and the proximity of livestock production facilities to homes, two factors that are currently potentially blocking the emergence of this type of project.

There is also a need to develop existing data on the quality of aquaponic products (health, nutritional and sensory), and on the social acceptability of this practice, which depends not only on consumer attitudes to the technique as such, but also on the environmental benefits it actually brings, which have not yet been demonstrated.

From an economic point of view, many existing systems, particularly in the United States, make little or no profit, and it does not seem that large-scale systems necessarily fare better than family-sized systems. There are, however, a growing number of examples of farms that have achieved a viable economic model, making it possible to invest and expand, which gives us good hope of seeing aquaponics develop further in the years to come. It will be interesting to see whether or not the companies quoted in this book are able to sustain their model and their economic success. Some choose to start with a modest size of operation that is not very dependent on technology, and then expand by duplicating the model, while others start from the outset with high-tech systems that are generally heavy in terms of investment and production costs: the choice of one or other of these entry points is obviously conditioned by the ability of the project leaders to provide self-financing or to convince investors. It is also worth mentioning models that have ended in financial failure: Farmed Here in the United States and Urban Farmers in Holland, for example, went bankrupt in 2017 and 2018 respectively after just a few years in business, even though millions of euros had been invested in these projects. But it's not easy to find reports analysing the real reasons for these failures, so we can only speculate. However, by following a few basic rules, you can limit the risk of failure: stand out from the crowd, position yourself in niche markets, sell locally, be cautious about financial assumptions, know how to communicate very well, have a technical perspective and know how to surround yourself with solid technical and regulatory skills.

Notes

[7] The species susceptible to the regulated diseases are specified in the Order of 29 July 2013 and the vector species for these diseases are specified in Annex I of Regulation (EC) 1251/2008.
[8] https://www.cultures-aquaponiques.com/ (consulted on 18/02/2019).

General conclusion

_____

Historically conceived and developed by aquaculture research in the 1970s and 1980s with a view to the phytodepuration of effluents, with no focus on adding value to these ancillary products, aquaponics is now of interest to the market garden/horticultural and fish farming sectors, which are looking for ideas to diversify and rethink their production systems. It is also attracting interest from the general public and from institutions that see it as a way of producing fish and plants together in a sustainable way, as part of a circular economy approach based on recycling water and inputs. The aquaponics approach requires us to think differently about previously isolated forms of production: what used to be regarded as waste (fish effluent) becomes a valuable resource.

From a technical point of view, the current body of knowledge on aquaponics contains a number of gaps and raises questions that need to be investigated further:

- how do you find the right balance between different species of fish, different types of feed and different species of plants? It is necessary to establish the proportion of macro- and micronutrients that fish can release into the water for a given feed in a given system; this depends both on fish species, fish density and temperature. The uptake and use of these nutrients will depend on the type of plants grown (leafy plants, fruiting plants) and the stage of cultivation. Numerous fish/plant pairs need to be tested in order to acquire ever more data;
- what is the real environmental impact of aquaponics? Lifecycle analyses are needed to assess the relevance of aquaponics from an environmental point of view, compared with traditional cultivation techniques (open field, hydroponics) and aquaculture (open circuit, closed circuit);
- How can we explain the excellent yield results obtained in aquaponics, when the work carried out over the years in the field of hydroponics would tend to suggest that aquaponics should not be very productive? In aquaponics, conductivity is very low, pH seems too high, mineral levels are too low and present in unfavourable ratios... and yet it works. The bacterial and fungal flora present in the fish ponds, biofilter and culture substrates probably have an important role to play. It is useful to define the existing interactions of these micro-organisms with plant growth. As part of the APIVA® project, work based on a metagenomic approach will be carried out to define the bacterial populations associated with major metabolic roles in the various compartments of the system;
- what are the nutritional and sensory qualities of plants grown in aquaponics? Productivity potential should not be the only argument in favour of the competitiveness of aquaponics

compared with hydroponics or open field cultivation. The APIVA® project has provided some answers on the description and quantification of nutrient flows in an aquaponics system and on the purification effect that plants can exert on fish farm waste. The work carried out has made it possible to understand and model the operation of systems based on defined fish/plant pairs, and thus to establish design elements that can be transferred to professionals.

From an economic and social point of view, the economic relevance of aquaponics also raises questions. A number of commercial structures have been set up since 2005 in the United States and Australia, and more recently in Europe, including France, but there are a number of major obstacles to the development of truly profitable systems: high initial investment, higher production costs than hydroponics due to the significant operational costs associated with energy and fish feed requirements. So should aquaponics be seen as a single system, where the fish are a source of fertiliser for the plants, without the fish compartment being profitable? Or should we be aiming for a different objective, which would be to achieve profitability on both fronts by combining two complementary forms of production that are sufficient on their own to be profitable? Another difficulty is that current regulations do not provide a framework for this new activity, and the best we can do is try to interpret existing regulations in such a way as to consider them applicable, by extension, to aquaponics. A legal framework seems necessary to define the methods of production and marketing of products from such systems, in order to anticipate societal questions about this practice.

Economic feasibility also concerns the problems of marketing products: the current impossibility of acquiring the organic label, and the scarcity of official labels enabling aquaponics to really stand out in terms of production quality and respect for the environment, raises questions about the possibility of being competitive with products from traditional intensive agriculture. Is this what we're looking for? Or do we want to offer fresh, locally-grown, quality products to reach a very specific market segment? It's worth noting that aquaponics is generally perceived positively by consumers, as long as you explain the principle, the origin and the quality of the products. However, it remains essential to study the extent to which consumers would be prepared to pay a higher price for products from such systems, which is *a* necessary condition for a viable economic model.

Finally, we note that the profiles of "aspiring aquaponiculturists" are varied, and include many new entrants to the aquaculture or plant industry, often people undergoing professional retraining and seeking to acquire the skills to successfully complete their project. To date, fish farmers, market gardeners and horticulturists by trade have been relatively under-represented among project initiators, a symptom of the difficulty of questioning existing practices, even if it is increasingly apparent that aquaponics is intriguing and interesting. In the years to come, it seems likely that this technique will be developed within existing structures, or even in association with strategies involving producers of fish and vegetable or ornamental plants.

Glossary

————————————

Aerobic: Refers to a living organism or mechanism that needs air or oxygen to function.

Allelochemical compounds: substances involved in communication between different species, as part of a phenomenon known as "allelopathy", which is a set of multiple direct or indirect, positive or negative biochemical interactions. These are usually interactions between one species of plant and another (including micro-organisms) by means of metabolites such as phenolic acids, flavonoids, terpenoids and alkaloids.

Anaerobic: Refers to a living organism or mechanism that does not require air or oxygen to function.

Apex: in botany, the apex is the end of a root or stem.

ATP: in the biochemistry of all known living organisms, ATP (adenosine triphosphate) provides the energy required for chemical reactions in metabolism, locomotion, cell division and the active transport of chemical species across biological membranes.

Autotrophy: the production of organic matter by a living organism through the reduction of inorganic and mineral matter. This mode of nutrition is characteristic of chlorophyllous (green) plants, cyanobacteria and sulphur bacteria.

Auxin: plant growth hormone essential for plant development.

Bioaccumulation: the ability of certain organisms (plants, animals, fungi, microbes) to absorb and concentrate in all or part of their bodies certain chemical substances that may be rare in the environment (useful or essential trace elements, or undesirable toxic substances).

Buffer capacity: quantity (expressed in number of moles) of acid or strong base to be added to 1 L of buffer solution to change the pH by one unit. The greater the buffer capacity, the less the pH of the buffer solution will change with the addition of an acid or base.

Cation exchange capacity (CEC): quantity of cations that a soil can retain on its adsorbent complex at a given pH. CEC is used as a measure of soil fertility, indicating the nutrient retention capacity of a given soil.

Chelate: chelation is a physico-chemical process during which a complex, the chelate, is formed between a ligand, known as the "chelator", and a metal cation (or atom), then complexed, known as the "chelated".

Chlorosis: wilting and yellowing of plants due to a lack of chlorophyll, caused by various types of deficiency in certain macro- or micronutrients.

Collar: part of the plant between the stem and the roots.

Dryness rate: percentage by mass of dry matter. For example, a sludge with a dryness of 10% has a moisture content of 90%.

Eutrophication: an imbalance in the aquatic environment caused by an increase in the concentration of nitrogen and phosphorus. It is characterised by excessive growth of plants and algae due to the high availability of nutrients.

Evapotranspiration: in plants, transpiration is the continuous process caused by the evaporation of water by the leaves following the uptake of water by the roots. Transpiration is the main driving force behind the circulation of sap and takes place mainly through the stomata.

Gravitropism: in plant physiology, gravitropism is the way in which plants develop and orient themselves in relation to gravity.

Haber process: chemical process used to synthesise ammonia (NH_3) by hydrogenating atmospheric nitrogen gas (N_2) with hydrogen gas (H_2) in the presence of a catalyst. The German chemist Fritz Haber perfected this chemical process in 1909. Ammonia is most often used to create synthetic nitrogen fertilisers, considered essential for feeding the world's population at the beginning of the 21st century.

Heterotrophy: the need for a living organism to feed on pre-existing organic components. Heterotrophy is the opposite of autotrophy.

Industrial fishing: fishing intended to supply industrial sectors with small pelagic fish of low commercial value, which are processed into fishmeal and fish oil used in various animal production sectors (particularly aquaculture) in the formulation of commercial feed.

Instream flow: compulsory minimum flow of water (expressed as a percentage of the total mean flow) that the owners or managers of a hydraulic structure (dam, weir, hydroelectric unit, etc.) must reserve for the watercourse and the minimum functioning of ecosystems. Since 1st January 2014, Law no. 2006-1772 of 30 December 2006 on water and aquatic environments has set the instream flow for fish farms at $1/10^e$ of the modulus of the watercourse. The withdrawable flow for a fish farmer is the difference between the total flow of a watercourse and its instream flow.

Internode: the space between two nodes or joints on a stem.

Interveinal: the area between the veins of a leaf.

Metagenomics: an approach which, through direct sequencing of the DNA present in the sample, provides a genomic description of the sample's content, as well as an insight into the functional potential of an environment.

Off-flavor: refers to an unpleasant "muddy" taste in fish products, caused by contamination of the water and fish flesh with odoriferous substances. The main metabolic products responsible for the muddy taste are geosmin and methylisoborneol. These two organic compounds are of microbial origin.

Organoleptic: refers to anything likely to excite a sensory receptor, such as appearance, smell, taste, texture or consistency. These indicators are used in organoleptic tests in the food industry to assess food quality.

Osmoregulation: the set of processes involved in regulating the concentration of salts dissolved in the internal fluids of living organisms.

Osmotic pressure: minimum pressure required to prevent the passage of a solvent from a less concentrated solution to a more concentrated solution through a semi-permeable membrane.

Phenological stage: a specific period in the growth or development of a plant, such as flowering, leafing, fruiting, leaf colouring or leaf fall.

Prophylaxis: all the measures to be taken to prevent disease.

Stomata: natural opening in the epidermis of the stem or leaf, which ensures gas exchange with the outside environment (respiration, excretion).

Bibliography

Articles, books and websites

Acierno R., Blancheton J.P., Bressani G., Ceruti L., Chadwick D., Roque d'Orbcastel E., Claricoates J., Donaldson G., Donaldson G., 2006. *Manual on effluent treatment in aquaculture: Science and Practice*, Aquaetreat, 163 p., https://www.researchgate.net/publication/29494856_Manual_on_effluent_treatment_in_aquaculture_Science_and_Practice_Aquaetreat/download (accessed 28/02/2019).

Adler P.R., 2000. Economic evaluation of hydroponics and other treatment options for phosphorus removal in aquaculture effluent. *Hortscience*, 35 (6), 993-999.

Adler P.R., 2000. Economic analysis of an aquaponic system for the integrated production of rainbow trout and plants. *International Journal of Recirculating Aquaculture*, 1, 15-34.

Aerts R., De Schutter B., Rombouts L., 2002. Suppression of Pythium spp. by Trichoderma spp. during germination of tomato seeds in soilless growing media. Meded. Gent Fak. Landbouwkd. Toegep. *Biologische Wetenschappen*, 67, 343-351.

Afsharipoor S., Roosta H.R., 2010. Effect of different planting beds on growth and development of strawberry in hydroponic and aquaponic cultivation systems. *Plant Ecophysiology*, 2, 61-66.

AGRESTE, 2013. Exploitations légumières, les surfaces. *AGRESTE : Les dossiers*, 16, 53-97.

Ahemad M., Kibret M., 2014. Mechanisms and applications of plant growth promoting rhizobacteria: Current perspective. *Journal of King Saud University-Science*, 26, 1-20.

Al Hafedh Y.S, Alam A., Beltagi M.S., 2008. Food production and water conservation in a recirculating aquaponic system in Saudi Arabia at differents ratios of fish feed to plants. *Journal of the World Aquaculture Society*, 39 (4), 510-520.

Anthonisen A.C., Loehr R.C., Prakasam T.B.S., Srinath E.G., 1976. Inhibition of nitrification by ammonia and nitrous acid. *Journal WPCF*, 48, 835-852.

Apiva : aquaponics, plant innovation and aquaculture, 2019. https://commissaires-priseuses/le-projet-apiva-objectifs-et-partenaires/ (accessed 12/03/2019)

Aquinove, 1995. *Study of micro-screening of fish farm effluents using a rotary drum filter*. Comparison of 6 models, operational characteristics, 28 p.

Bailey D.S., Rakocy J.E., Cole W.M., Shultz K.A., 1997. Economic analysis of a commercial-scale aquaponic system for the production of tilapia and lettuce. p. 603-612, *In*: Tilapia Aquaculture: Proceedings from the Fourth International Symposium on Tilapia in Aquaculture 1997 (Fitzsimmons, K.), NRAES, Ithaca, N.Y., 808 p.

Bailey D.S., Ferrarezi R.S., 2017. Valuation of vegetable crops produced in the UVI Commercial Aquaponic System. *Aquaculture Reports*, 7, 77-82.

Bajsa O., Nair J., Mathew K., Ho G.E., 2003. Vermiculture as a tool for domestic wastewater management. *Water Science and Technology*, 48, 125-132.

Bartelme R.P., Oyserman B.O., Blom J.E., Sepulveda-Villet O.J., Newton R.J., 2018. Stripping away the soil: plant growth promoting microbiology opportunities in aquaponics. *Frontiers in Microbiology*, 9 (8), 1-7.

Belaud D., 1996. *Oxygenation de l'eau en aquaculture intensive*, éditions Cépaduès, Toulouse, 208 p.

Bergheim A., Cripps S.J., Liltved H., 1998. A system for the treatment of sludge from land-based fish-farms. *Aquatic Living Resources*, 11 (4), 279-287.

Bergheim A., Brinker A., 2003. Effluent treatment for flow-through systems and European environmental regulations. *Aquaculture Engeneering*, 27, 61-77.

Berthelot C., Leyval C., Foulon J., 2016. Plant growth promotion, metabolite production and metal tolerance of dark septate endophytes isolated from metal-polluted poplar phytomanagement sites. *FEMS Microbiology Ecology*, 92, fiw144. doi: 10.1093/femsec/fiw144.

Betancor M.B., Sprague M., Sayanova, O. Usher, S. Campbell, P.J. Napier, J.A. Caballero M.J., Tocher D.R., 2015. Evaluation of a high-EPA oil from transgenic Camelina sativa in feeds for Atlantic salmon (*Salmo salar* L.): Effects on tissue fatty acid composition, histology and gene expression. *Aquaculture*, 444, 1-12.

Bittsansky A., Pilinszki K., Gyulai G., Komives T., 2015. Overcoming ammonium toxicity. *Plant Science*, 231, 184-190.

Bittsansky A., Uzinger N., Gyulai G., Mathis A., Junge R., Villarroel M., Kotzen B., Komives T., 2016. Nutrient supply of plants in aquaponic systems. *Ecocycles*, 2 (2), 17-20.

Blancheton J.P., Dosdat A., Deslous Paoli, J.M., 2004. Minimisation of biological discharges from fish farms. *Dossiers de l'environnement de l'Inra*, 26, 67-78.

Blancheton J.P., Attramadal K.J.K., Michaud L., d'Orbcastel, E.R., Vadstein O., 2013. Insight into bacterial population in aquaculture systems and its implication. *Aquaculture engineering,* 53, 30-39.

Blidariu F., Grozea A., 2011. Increasing the economical efficiency and sustainability of indoor fish farming by means of aquaponics - Review. *Animal Science and Biotechnologies*, 44 (2), 8.

Blidariu F., Radulov I., Lalescu D., Drasovean A., Grozea A., 2013. Evaluation of nitrate level in green lettuce conventional grown under natural conditions and aquaponic system. *Animal Science and Biotechnologies*, 46 (1), 244-250.

Boulard T., Brun R., Jaffrin A., Jeannequin B., 1999. *Fertirrigation et recyclage des solutions nutritives en cultures hors-sol sous abris*, ADEME, 28 p.

Bosma R.H., Lacambra L., Landstra Y., Perini C., Poulie J., Schwaner M.J., Yin Y., 2017. The financial feasibility of producing fish and vegetables through aquaponics. *Aquacultural Engineering*, 78, Part B, 146-154.

Boxman S.E., Zhang Q., Bailey D., Trotz M.A., 2016. Life cycle assessment a commercial-scale freshwater aquaponic system. *Environmental Engineering Science*, 34 (5), 299-311.

Bron G., 2012. *L'entreprise horticole : approche globale et durable*, educagri éditions, Dijon, 392 p.

Buzby K.M., Waterland N.L., Semmens K.J., Lin L-S., 2016. Evaluating aquaponic crops in a freshwater flow-through fish culture system. *Aquaculture*, 460, 15-24.

Caruso D., Devic E., Subamia I.W., Talamond P., Baras E., 2014. *Technical handbook of domestication and production of diptera Black Soldier Fly (BSF)* Hermetia illucens, Stratiomyidae, IRD éditions, Bondy, 159 p.

Centre d'études pour le développement d'une pisciculture autonome, 2013. http://cedepa.fr/pisciculture-perspectives/ (accessed 28/03/2019).

Chaves P.A., Sutherland R.M., Larid L.M., 2008. An economic and technical evaluation of integrating hydroponics in a recirculation fish production system. *Aquaculture Economics and Management*, 3 (1), 83-91.

Chopin T., Robinson S.M.C., Troell M., Neori A., Buschmann A., Fang J.G., 2008. Multitrophic integration for sustainable marine aquaculture. In: *Encyclopedia of Ecology* (Sven E. *et al.*), Academic Press, Oxford, 2463-2475.

Cockx E., Simonne E.H., 2003. *Reduction of the impact of fertilization and irrigation on processes in the nitrogen cycle in vegetable fields with BMPs*, Horticultural Sciences Publication, University of Florida, HS948, 17 p., http://edis.ifas.ufl.edu/pdffiles/HS/HS20100.pdf (accessed 28/02/2019).

Connolly K., Trebic T., 2010. Optimization of a backyard aquaponic food production system. Faculty of Agricultural and Environmental Sciences Macdonald Campus, McGill University, 74 p., https://www.mcgill.ca/bioeng/files/bioeng/KeithTatjana2010.pdf (accessed 28/03/2019).

Cook S.M., Zhan Z.R., Pickett J.A., 2007. The use of push-pull strategies in integrated pest management. *Annual Review of Entomology*, 52, 375-400.

Craig S., Helfrich L.A., 2009. Understanding fish nutrition, feeds, and feeding. Virginia Cooperative Extension, Virginia State University. Publication 420-268, 4 p.

Cultures Aquaponiques M.L. Inc, 2017. https://www.cultures-aquaponiques.com/ (accessed 18/02/2019).

Da Silva Cerozi B.S., Fitzsimmons K., 2016a. Use of *Bacillus* spp. to enhance phosphorus availability and serve as a plant growth promoter in aquaponics systems. *Scientia Horticulturae*, 211, 277-282.

Da Silva Cerozi B.S., Fitzsimmons K., 2016b. The effect of pH on phosphorus availability and speciation in an aquaponics nutrient solution, *Bioresource Technology*, 219, 4, doi:10.1016/j.biortech.2016.08.079.

Da Silva Cerozi B.S., Fitzsimmons K., 2016c. Phosphorus dynamics modeling and mass balance in an aquaponics system. *Agricultural Systems*, 153, 94-100.

Dalsgaard J., Pedersen, P.B., 2011. Solid and suspended/dissolved waste (N, P, O) from rainbow trout (*Oncorynchus mykiss*). *Aquaculture*, 313, 92-99, doi: .10.1016/j.aquaculture

Danaher J.J., Schultz R.C., Rakocy J.E., Bailey D.S., 2013. Alternative solids removal for warm water recirculating raft aquaponics systems. *Journal of the world aquaculture society*, 44 (3), 374-383.

Damon E., Seawright R.B., Walker R.R.S., 1998. Nutrient dynamics in integrated aquaculture hydroponics systems. *Aquaculture*, 160, 215-237.

Davidson J., Good C., Welsh C., Summerfelt S., 2011. The effects of ozone and water exchange rates on water quality and rainbow trout performance in replicated water recirculating systems, *Aquacultural Engineering*, 44, 80-96.

Davidson J., Good C., Welsh C., Summerfelt S., 2014. Comparing the effects of high vs. low nitrate on the health, performance, and welfare of juvenile rainbow trout *Oncorhynchus mykiss* within water recirculating aquaculture systems, *Aquaculture Engineering*, 59, 30-40.

Dediu L., Cristea V., Docan A., Vasilean I., 2011. Evaluation of condition and technological performance of hybrid bester reared in standard and aquaponic system. *AACL Bioflux*, 4 (4), 490-498.

Dediu L., Cristea V. Xiaoshuan Z., 2012. Waste production and valorization in an integrated aquaponic system with bester and lettuce. *African Journal of Biotechnology*, 11 (9), 2349-2358.

Delaide B., Goddek S., Gott J., Soyeurt H., Jijakli H., 2016. Lettuce (*Lactuca sativa* L. var. Sucrine) growth performance in complemented aquaponic solution outperforms hydroponics. *Water*, 8, 467, doi:10.3390/w8100467.

Delaide B., Delhaye G., Dermience M., Gott J., Soyeurt H., Jojalki H., 2017. Plant and fish production performance, nutrient mass balances, energy and water use of the PAFF Box, a small-scale aquaponic system. *Aquacultural Engineering*, 78, Part B, 130-139.

Directions départementales des Territoires et de la mer, 2016. Agricultural facilities. Guide technique pour l'instruction des autorisations d'urbanisme. Bretagne, 11 p., http://www.aveyron.chambagri.fr/fileadmin/documents_ca12/Aveyron/EspaceAgriculteur/Extrait_r%C3%A8gles_d_urbanisme_et_constructions_agricoles.pdf (accessed 20/03/2019).

Diver S., 2006. Aquaponics-Integration hydroponics with aquaculture. A publication of ATTRA - National Sustainable Agriculture Information Service, 28 p, https://attra.ncat.org/attra-pub/download.php?id=56 (accessed 28/02/2019).

Dolomatov S., Zukow W., Hanger-Derengowska M., Kozestanksa M., Jaorska I., Nalazek, 2013. Toxic and physiological aspects of metabolism of nitrites and nitrates in the fish organism. *Journal of Health Sciences*, 3 (2), 68-91.

Dosdat A., Servais F., Mctailler R., Huelvan C., Desbruykres E., 1996. Comparison of nitrogenous losses in five teleost fish species. *Aquaculture*, 141, 107-127.

Ebeling J.M., Timmons M.B., 2006. Engineering analysis of the stoichiometry of photoautotrophic, autotrophic, and heterotrophic removal of ammonia-nitrogen in aquaculture systems, *Aquaculture*, 257, 346-358.

Eding E.H., Kamstra A., Verreth J.A.J., Huisman E.A., Klapwijk A., 2006. Design and operation of nitrifying trickling filters in recirculating aquaculture: A review. *Aquacultural Engineering*, 34, 234-260.

Edwards P., 2015. Aquaculture environment interactions: past, present and likely future trends. *Aquaculture*, 447, 2-14.

El Komy M.H, Saleh A.A., Eranthodi A., Molan Y.Y., 2015. Characterization of novel *Trichoderma asperellum* isolates to select effective biocontrol agents against tomato fusarium, *Plant Pathology Journal*, 31 (1), 50-60.

Elumalai S.A., Shaw A.M., Pattillo D.A., Currey C.J., Rosentrater K.A., Xie K., 2017. Influence of UV treatment on the food safety status of a model aquaponic system. *Water*, 9, 27, doi:10.3390/w9010027.

Emerson K., Russo R.C., Lund R.E., Thurston R.V., 1975. Aqueous ammonia equilibrium calculations: effect of pH and temperature. *Journal of the Fisheries Research Board of Canada*, 32 (12), 2379-2383.

Endut A., Jusoh A., 2009. Effect of flow rate on water quality parameters and plant growth of water spinach (*Ipomea aquatica*) in an aquatic recirculating system. *Desalination and Water Treatment*, 5 (1-3), 19-28, doi:10.5004/dwt.2009.559.

Endut A., Jusoh A., Ali N., Nik W.B.W., Hassan A., 2010. A study on the optimal hydraulic loading rate and plant ratios in recirculating aquaponic system. *Bioresource Technology*, 101, 1511-17.

Endut A., Jusoh A., Ali N., Nik W.B.W., 2011. Nutrient removal from aquaculture wastewater by vegetable production in aquaponics recirculation system. *Desalination and Water Treatment*, 32 (1-3), 422-430.

Endut A., Jusoh A., Ali N., 2014. Nitrogen budget and effluent nitrogen components in aquaponics recirculation system. *Desalination and Water Treatment*, 52 (4-6), 744-752.

Engle C.R., 2015. *Economics of aquaponics*. SRAC Publication, 5006, 4 p.

Ernst G., Emmerling C., 2009. Impact of five different tillage systems on soil organic carbon content and the density, biomass and community composition of earthworms after a ten-year period. *European Journal of Soil* Biology, 45 (3), 247-251, doi:10.1016/j.ejsobi.2009.02.002.

Espinosa Moya E.A., Angel Sahagun C.A., Mendoza Carrillo J.M., Albertos Alpuche P.J., 2014. Herbaceous plants as part of biological filter for aquaponics system. *Aquaculture research*, 1-11.

EU Aquaponics Hub, 2019, https://euaquaponicshub.com/eu-aquaponics-map/ (accessed 12/03/2019).

Fang Y., Hu Z., Zou Y., Zhang J., Zhu Z., Zhang J., Nie L., 2017a. Improving nitrogen utilization efficiency of aquaponics by introducing algal-bacterial consortia. *Bioresource Technology*, 245, 358-364.

Fang Y., Hu Z., Zou Y., Fan J., Wang Q., Zhu Z., 2017b. Increasing economic and environmental benefits of media-based aquaponics through optimizing aeration pattern. *Journal of Cleaner Production*, 162, September, 1111-1117.

FAO, 2018. The state of world fisheries and aquaculture. Contributing to food security and nutrition for all, Rome, 254 p., http://www.fao.org/3/i9540fr/i9540fr.pdf (accessed 19/02/2019).

Forchino A.A., Lourguioui H., Brigolin D., Pastres R., 2017. Aquaponics and sustainability: the comparison of two different aquaponic techniques using the Life Cycle Assessment (LCA). *Aquacultural Engineering*, 77, 80-88.

Fox B.K., Tamaru C., Radovich T., Klinger-Bowen R., MCGovern-Hopkins K., Bright L., Pant A., Gurr I., Sugano J., Brent S., Lee C.N., 2011. Beneficial use of vermicompost in aquaponics vegetable production. *Hanai'Ai, The Food Provider*, 10, 1-5.

Fox B.K, Tamaru C.S, Hollyer J., Castro L.F, Monseca J.M, Jay-Russell M., Low Todd, 2012. A preliminary study of microbial water quality related to food safety in recirculating aquaponic fish and vegetable production systems. *Food Safety and Technology*, 51, 1-11.

France Agrimer, 2013. Observatoire structurel des entreprises de production de l'horticulture et de la pépinière ornementales. Synthèse France, [online], 24 p., <http://www.franceagrimer.fr/content/download/26855/235427/file/ETU-HOR-2013-ObsStructEnt-SynthFRANCE2012.pdf> (accessed 28/02/2019).

Fredricks K.T., 2015. Literature review of the potential effects of formalin on nitrogen oxidation efficiency of the biofilters of recirculating aquaculture systems (RAS) for freshwater finfish. U.S. Geological Survey (USGS), Reston, Virginia, Report 2015-1097, 26 p., https://pubs.usgs.gov/of/2015/1097/pdf/ofr2015-1097.pdf (accessed 28/02/2019).

Fujiwara K., Lida Y., Iwai T., Aoyama C., Inukai R., Ando A., Ogawa J., Ohnishi J., Terami F., Takano M., Shinohara M., 2013. The rhizosphere microbial community in a multiple parallel mineralization system suppresses the pathogenic fungus *Fusarium oxysporum*. *Microbiologyopen*, 2 (6), 997-1009, doi:10.1002/mbo3.140.

Ghaly A.E., Kamal M., Mahmoud N.S., 2005. Phytoremediation of aquaculture wastewater. *Environment International*, 31, 1-13.

Ghosh B.P., Burris R.H., 1950. Utilization of nitrogenous compounds by plants. *Soil Science*, 70 (3), 187-204.

Gislerød H.R., Adams P., 1983. Diurnal variations in the oxygen content and requirement of recirculating nutrient solutions and in the uptake of water and potassium by cucumber and tomato plants. *Scientia Horticultura*, 21 (4), 311-321.

Goddek S., Delaide B., Mankasingh U., Ragnasdottir K.V., Jijalki H., Thorarinsdottir R., 2015. Challenges of sustainable and commercial aquaponics. *Sustainability*, 7, 4199-4224.

Goddek S., Espinal C.A., Delaide B., Jijakli M.H., Schmautz Z., Wuertz S., Keesman J., 2016a. Navigating towards decoupled aquaponic systems: a system dynamics design approach. *Water*, 8, 303, doi:10.3390/w8070303.

Goddek S., Schmautz Z., Scott B., Delaide B., Wuertz S., Junge R., 2016b. The effect of anaerobic and aerobic fish sludge supernatant on hydroponic lettuce. *Agronomy*, 6, 37.

Graber A., Junge R., 2009. Aquaponics systems: nutrient recycling from fish wastewater by vegetable production. *Desalination*, 246, 147-156.

Gravel V., Martinez C., Antoun H., Tweddell R.J., 2006. Control of greenhouse tomato root rot (*Pythium ultimum*) in hydroponic systems, using plant-growth-promoting microorganisms. *Canadian Journal of Plant Pathology*, 28, 475-483, doi:10.1080/07060660609507322.

Gravel V., Dorais M., Vandenberg G., 2015. Fish effluents promote root growth and suppress fungal diseases in tomato transplants. *Canadian Journal of Plant Pathology*, 95 (2), 427-436.

Gruda N., 2008. Do soilless culture systems have an influence on product quality of vegetables? *Journal of Applied Botany and Food Quality*, 82, 141-147.

Guillaume J., Kaushik S., Bergot P., Métailler R., 1999. *Nutrition et alimentation des poissons et crustacés*, Inra/Ifremer, Paris/Brest, 490 p.

Heidemann K., 2015a. Economic Analysis of Commercial Aquaponic Production Systems, *Sustainable Agriculture Research & Education*, College Park, https://projects.sare.org/project-reports/gs13-125/ (accessed 28/02/2019).

Heidemann K., 2015b. Commercial aquaponics case study #3: economic analysis of the University of the Virgin Islands commercial aquaponics system, Department of agricultural economics, University of Kentucky, College Park, 11 pp, https://www.uky.edu/Ag/AgEcon/pubs/extaec2015-1821.pdf (accessed 28/02/2019).

Heidemann K., 2015c. Commercial aquaponics case study #1: economic analysis of Lily Pad Farms, Department of agricultural economics, University of Kentucky, College Park, 11 p., https://www.uky.edu/Ag/AgEcon/pubs/extaec2015-0330.pdf (accessed 28/02/2019).

Heidemann K., 2015d. Commercial aquaponics case study #2: economic analysis of Traders Hill Farms, Department of agricultural economics, University of Kentucky, College Park, 8 p., https://www.uky.edu/Ag/AgEcon/pubs/extaec2015-0426.pdf (accessed 28/02/2019).

Hochmuth G.J., 2012. Fertilizer management for greenhouse vegetables - Florida Greenhouse Vegetable Production Handbook. Horticultural Sciences Department, Florida Cooperative Extension Service, Institute of Food and Agricultural Sciences, University of Florida, 3, 19 p.

Hoevenaars K., Junge R., Bardocz T., Leskovec M., 2018. EU policies: New opportunities for aquaponics. *Ecocycles*, 4 (1), 10-15, doi:10.19040/ecocycles.v4i1.87.

Hopkins W.G., 2003. *Plant Physiology*, De Boeck, Louvain-la-Neuve (Belgium), 514 p.

Hollyer J., Tamaru C., Riggs A., Klinger-Bowen R., Howerton R., Okimoto D., Castro L., Ron B.R., Troegner V., Martinez Glenn, 2009. On-farm food safety: aquaponics. *Food Safety and Technology*, 38, 1-7.

Hosseinzadeh S., Bonarrigo G., Verheust Y., Roccaro P., van Hulle S., 2017. Water reuse in closed hydroponic systems: comparison of GAC adsorption, ion exchange and ozonation processes to treat recycled nutrient solution. *Aquacultural Engineering*, 78, 190-195, doi:10.1016/j.aquaeng.2017.07.007.

Hrubec T.C, 1996. Nitrate toxicity a potential problem of recirculating systems. *Aquaculture Engineering Society Proceedings*, 41-48.

Hu, Z., Lee, J.W., Chandran, K., Kim, S., Brotto, A.C., Khanal, S.K., 2015. Effect of plant species on nitrogen recovery in aquaponics, *Bioresource Technology*, 188, 92-98.

Ibiene A.A., Agogbua J.U., Okonko I.I., Nwachi G.N., 2012. Plant growth promoting rhizobacteria (PGPR) as biofertilizer: effect on growth of *Lycopersicum esculentus*. *Journal of American Science*, 8 (2), 318-324.

Joly A., Junge R., Bardocz T., 2015. Aquaponics business in Europe: some legal obstacles and solutions. *Ecocycles*, 1 (2), 3-5.

Jones J.B., 2005. *Hydroponics: a practical guide for the soilless grower*, CRC Press Online, Boca Raton (Florida), 440 p.

Jones J.B., 2014. *Complete Guide for growing plants hydroponically*, CRC Press Online, Boca Raton, Florida, 206 pp.

Jung I.S., Lovitt R.W., 2011. Leaching techniques to remove metals and potentially hazardous nutrients from trout farm sludge. *Water Res*, 45, 5977-5986.

Junge R., König B., Villarroell M., Komives T., Jijakli M.H., 2017. Strategic Points in Aquaponics. *Water*, 9, 182, doi:10.3390/w9030182.

Kane C.D., Jasoni R.L., Peffley E.P., Thompson L.D., Green C.J., Pare P., Tissue D., 2006. Nutrient solution and solution pH influences on onion growth and mineral content. *Journal of Plant Nutrition*, 29, 375-390.

Kantartzi S.G., Vaiopoulou E., Kapagiannidis A., Aivasidis, 2006. Kinetic characterization of nitrifying pure cultures in a chemostat. *Global NEST Journal*, 8, 43-51.

Karthikeyan V., Gopalakrishnan A., 2014. A novel report of phytopathogenic fungi *Gilbertella persicaria* infection on *Penaeus monodon*. *Aquaculture*, 430, 224-229.

Kaushik S., 1990. Fish nutrition and waste control. *Pisciculture Francaise*, 101, 14-23.

Kaushik S.J., 1998a. Nutritional bioenergetics and estimation of waste production in non salmonids. *Aquatic Living Resources*, 11, 211-217.

Kaushik S., Guillaume J., Bergot P., Métailler R., 1999, Nutrition et alimentation des poissons et des crustacés, Inra éditions/Ifremer, Versailles/Brest, 492 p.

Keeratiurai P., 2013. Efficiency of wastewater treatment with hydroponics. *ARPN. Journal of Agricultural and Biological Science*, 8 (12), 800-805.

Kinkelin P., Michel C., Ghittino P., 1985. *Précis de pathologie des poissons*, Inra, Paris, 348 p.

Kiraly K.A., Pilinszky K., Bittsanszky A., Gyulai G., Komives T., 2013. Importance of ammonia detoxification by plants in phytoremediation and aquaponics. *Novenytermeles (Plant Production)*, 62, 99-102.

Klinger D., Naylor R., 2012. Searching for solutions in aquaculture: charting a sustainable course. *Annual Review of Environment and Resources*, 37, 247-76.

Kroupova H., Machova J., Svobodova Z., 2005. Nitrite influence on fish: a review. *Veterinarni Medicina*, 50 (11), 461-471.

Labbé L., Lefèvre F., Bugeon J.,Fostier A., Jamin M., Gaumé M., 2014. Design of an innovative recirculating water trout production system. *INRA Productions Animales*, 27 (2), 135-146.

Lam S.S, Ma N.L., Jusoh A., Ambak M.A., 2015. Biological nutrient removal by recirculating aquaponic system: Optimization of the dimension ratio between the hydroponic and rearing tank components. *International Biodeterioration and Biodegradation*, 102, 107-115.

Lee S., Lee J., 2015. Beneficial bacteria and fungi in hydroponic systems: types and characteristics of hydroponic food production methods. *Scientia Horticulturae*, 195, 206-215, doi:10.1016/j.scienta. 2015.09.011.

Lennard W.A., Leonard B.V., 2006. A comparison of three different hydroponic subsystems (gravel bed, floating and nutrient film technique) in an aquaponic test system. *Aquaculture International*, 14, 539-550.

Lennard W., 2012. Aquaponic system design parameters, solids filtration, treatment and re-use. Aquaponic Fact Sheet Series, 10 p., https://www.aquaponic.com.au/Solids%20filtration.pdf (accessed 28/03/2019).

Lennard W.A., 2015. Aquaponics: a Nutrient Dynamic Process and the Relationship to Fish Feeds. *World aquaculture,* September, https://www.was.org/articles/Aquaponics-Nutrient-Dynamic-Process-Relationship-to-Fish-Feeds.aspx#.XHfsQbhCeUl (accessed 28/02/2019).

Lennard W.A., 2018. *Commercial aquaponic systems, integrating fish culture with hydroponic plant production*. Aquaponic Solutions, Victoria, Australia, 375 pp.

Letard M., Erard P., Jeannequin B., 1995. *Maîtrise de l'irrigation fertilisante : tomate sous serre et abris, en sol et hors-sol*, CTIFL, Paris, 221 p.

Lewis W.M., Yopp J.H., Schramm H.L., Brandenburg A.M., 1978. Use of hydroponics to maintain quality of recirculated water in a fish culture system. Transaction of the american fisheries society, 107 (1), 92-99.

Liang J-Y., Chien Y-H., 2013. Effects of feeding frequency and photoperiod on water quality and crop production in a tilapia-water spinach raft aquaponics system. *International Biodeterioration and Biodegradation*, 85, 693-700.

Liang J-Y., Chien Y-H., 2015. Effects of photosynthetic photon flux density and photoperiod on water quality and crop production in a loach (*Misgurnus anguillicandatus*) - nest fern (*Asplenium nidus*) raft aquaponics system. International Biodeterioration and Biodegradation, 102, 214-222.

Loper S., 2014. Diagnosing nutrient deficiencies quick-reference. College of agriculture and life sciences cooperative extension. The University of Arizona, 2 p., https://www.azlca.com/uploads/documents/quick-guide-def.pdf (accessed 28/02/2019).

Lorena S. Cristea V., Oprea L., 2008. Nutrients dynamic in an aquponic recirculating system for sturgeon and lettuce (*Lactuca sativa*) production. *Zootehnie și Biotehnologii*, 41 (2), 137-143.

Losordo T.M., Masser M.P., Rakocy J., 1998. Recirculating aquaculture tank production systems: an overview of critical considerations. SRAC Publication, 451, 1-6, http://fisheries.tamu.edu/files/2013/09/SRAC-Publication-No.-451-Recirculating-Aquaculture-Tank-Production-Systems-An-Overview-of-Critical-Considerations.pdf (accessed 19/02/2019).

Losordo T.M., Westers H., 1994. *System carrying capacity and flow estimation. In: Aquaculture water reuse systems: engineering design and management* (Timmons, M.B., Losordo, T.M.), Elsevier, Amsterdam (The Netherlands), 9-60.

Love D.C., Fry J.P., Genello L., Hill E.S., Frederick J.A., 2014. An international survey of aquaponics practitioners. *PLoS ONE*, 9 (7), 67-74, doi:10.1371/journal.pone.0102662.

Love D.C., Fry J.P., Genello L., Hill E.S., Frederick J.A., 2014. Commercial aquaponics production and profitability: findings from an international survey. *Aquaculture*, 435, 67-74.

Love D.C., Uhl M.S., Genello L., 2015. Energy and Water Use of a Small-Scale Raft Aquaponics System in Baltimore, Maryland, United States. *Aquacultural Engineering*, 68, 19-27.

Ma Y., Prasad M.N.V., Rajkumar M., Freitas H., 2012. Plant growth promoting rhizobacteria and endophytes accelerate phytoremediation of metalliferous soils. *Biotechnology Advances*, 29 (2), 248-58.

Mangmang J.S., Deaker R., Rogers G., 2015. Maximising fish effluent utilisation for vegetable seedling production by Azospirillum brasilense. *Procedia Environmental Sciences*, 29, 179.

Marcotte D., 2007. Evaluation and optimisation of the performance of fish farm waste treatment technologies. Technology transfer document, No. 2007.2, SORDAC.

Mariscal-Lagarda M., Paez-Osuna F., Esquer-Méndez J.L., Guerrero-Monroy I., Romo del Vivar A., Félix-Gastelum R., 2012. Integrated culture of white shrimp (*Litopenaeus vannamei*) and tomato (*Lycopersicon esculentum Mill*) with low salinity groundwater: management and production. *Aquaculture*, 366-367, 76-84.

Marschner P., 2008. *Mineral Nutrition of Higher Plants*. Academic Press, 3rd Revised Edition, Elsevier, Cambridge, Massachusetts (USA), 672 p.

Martinie-Cousty E., Prévot-Madère J., 2017. Marine and continental aquaculture farms: issues and conditions for successful sustainable development. Les avis du CESE (Conseil économique, social et environnemental), 99 p., https://www.actu-environnement.com/media/pdf/news-29196-rapport-cese-fermes-aquacoles.pdf (accessed 28/03/2019).

Martins C.I.M., Eding E.H, Verdegem M.C.J., Heinsbroek L.T.N., Blancheton J.P., Roque d'Orbcastel E., Verreth J.A.J., 2010. New developments in recirculating aquaculture systems in Europe: a perspective on environmental sustainability. *Aquaculture Engineering*, 43 (3), 83-93.

Marton E., 2008. Polycultures of fishes in aquaponics and recirculating aquaculture. *Aquaponics Journal*, 48, 28-33.

Matia L., Rauret G., Rubio R., 1991. Redox potential measurement in natural waters. *Fresenius Journal of analytical Chemistry*, 339 (7), 455-462.

Mattson N., Merrill T., 2015. Symptoms of common nutrient deficiencies in hydroponic lettuce, E-Gro Research Update, Cornell University, #2015.09, http://egrouni.com/pdf/Mattson_Lettuce_2015_9.pdf (accessed 28/02/2019).

Mattson N., Merrill T., 2015. Symptoms of common nutrient deficiencies in hydroponic basil, E-Gro Research Update, Cornell University, #2016.04, http://www.e-gro.org/pdf/2016-4.pdf (accessed 28/02/2019).

Maucieri C., Nicoletto C., Junge R., Schmautz Z., Sambo P., Borin M., 2017. Hydroponic systems and water management in aquaponics, a review. *Italian Journal of Agronomy*, 3 (1), 1-11, doi: .10.4081/ija.2017.1012

Mc Intyre A., 2014. The future of Europe's horticulture sector - Strategies for growth. European public health alliance, https://epha.org/the-future-of-europes-horticulture-sector-strategies-for-growth/ (accessed 28/02/2019).

Mc Murtry, M.R., Sanders, D.C., Cure, J.D., Hodson, R.G., Haning, B.C., Amand, P.C.S., 1997. Efficiency of water use of an integrated fish/vegetable co-culture system. *Journal of the World Aquaculture Society*, 28, 420-428.

Miličić V., Thorarinsdottir R., Dos Santos M., Turnsek Hancic M., 2017. Commercial aquaponics approaching the european market: to consumers' perceptions of aquaponics products in Europe. *Water*, 9, 80, doi:10.3390/w9020080.

Monsees H., Kloas W., Wuertz S., 2017a. Decoupled systems on trial: eliminating bottlenecks to improve aquaponic processes. *PloS One*, 28, 1-18.

Monsees H., Keitel J., Paul M., Kloas W., Wuertz S., 2017b. Potential of aquacultural sludge treatment for aquaponics: evaluation of nutrient mobilization under aerobic and anaerobic conditions. *Aquaculture Environment Interactions*, 9, 9-18.

Morel P., Poncet L., Rivière M., 2000. *Un point sur les supports de culture horticoles*, Inra éditions, Paris, 90 p.

Morgan L., 2003. Hydroponic Sustrates. *The Growing Edge*, 15 (2), 54-56.

Morard P., 2005. Les cultures végétales hors-sol. Publications Agricoles Agen, April 1995, 304 p.

Munguia-Fragozo P., Alatorre-Jacome O., Rico-Garcia E., Torres-Pacheco I., Cruz-Hernandez A., Ocampo-Velasquez R., Garcia Trejo J.J., Guevara-Gonzalez G., 2015. Perspective for aquaponic systems: "omic" technologies for microbial community analysis. *BioMed Research International*, ID 480386, 1-10.

Naegel LCA, 1977. Combined production of fish and plants in recirculating water. *Aquaculture*, 10 (1), 17-24, https://doi.org/10.1016/0044-8486(77)90029-1 (accessed 19/02/2019).

Nalazek A., 2013. Toxic and physiological aspects of metabolism of nitrites and nitrates in the fish organism. *Journal of Health Sciences*, 3 (2), 68-91.

Nelson J.A., Bugbee B., 2014. Economic analysis of greenhouse lighting: light emitting diodes vs. high intensity discharge fixtures. *PloS One*, 9 (6).

Nemethy S., Bittsansky A., Schmautz Z., Junge R., Komives T., 2016. Protecting plants from pests and diseases in aquaponic systems. *Ecocycles*, 2 (2), 17-20.

Nozzi V., Graber A., Schmautz Z., Mathis A., Junge R., 2018. Nutrient management in aquaponics: comparison of three approaches for cultivating lettuce, mint and mushroom Herb. *Agronomy*, 8, 27, doi:10.3390/agronomy8030027.

Osvalde A., 2011. Optimization of plant mineral nutrition revisited: the roles of plant requirements, nutrient interactions, and soil properties in fertilization management. *Environmental and experimental biology*, 9, 1-8.

Pajand Z.O., Soltani M., Bahmani M., Kamali A., 2017. The role of polychaete Nereis diversicolor in bioremediation of wastewater and its growth performance and fatty acid composition in an integrated culture system with Huso huso. *Aquaculture Research*, 48 (10), 1-9.

Palm H.W., Seidemann R., Wehofsky S., Knaus U., 2014a. Significant factors affecting the economic sustainability of closed aquaponic systems. Part I: System design, chemo-physical parameters and general aspects. *AACL Bioflux*, 7 (1), 20-32.

Palm H.W., Bissa K., Knaus U., 2014b. Significant factors affecting the economic sustainability of closed aquaponic systems. Part II: Fish and plant growth. *AACL Bioflux*, 7 (3), 162-175.

Palm H.W., Nievel M., Knaus U., 2014c. Significant factors affecting the economic sustainability of closed aquaponic systems. Part III: Plant units. *AACL Bioflux*, 8 (1), 89-106.

Pantanella E., Cardarelli M., Colla G., Rea E., Marcucci A., 2010. Aquaponics vs hydroponics: production and quality of lettuce crop. *Acta horticulturae*, 927, 887-894.

Pantanella E., Danaher J.J., Rakocy J.E., Shultz R.C., Bailey D.S., 2011a. Alternative media types for seedling production of lettuce and basil. *Acta Horticulturae,* 891, 257-264.

Pantanella E., Cardarelli M., Colla G., Rea E., Marcucci A. 2011b. Aquaponics vs hydroponics: production and quality of lettuce crop. *Acta Horticulturae*, 927, 887-893.

Pantanella E., Cardarelli M., Colla G., 2012. Yields and nutrient uptake from three aquaponic sub-systems (floating, NFT and substrate) under two different protein diets. *In* Proceedings, AQUA2012, Global Aquaculture securing our future. Prague, Czech Republic.

Pantanella E., Cardarelli M., Colla G. Di Mattia E., 2015. Aquaponics and food safety: effects of UV sterilization on total coliforms and lettuce production. *Acta Horticulturae*, 1062, 71-76.

Papatryphon E., Petit J., Van Der Werf H.M.J., Kaushik J.S., Kanyarushoki C., 2005. Nutrient-balance modeling as a tool for environmental management in aquaculture: the case of trout farming in France. *Environmental Management*, 35 (2), 161-174.

Petrea S.M., Cristea V., Dediu L., Contoman M., Lupoae P., Mocanu M., Coada T., 2013. Vegetable production in an integrated aquaponic system with rainbow trout and spinach. *Bulletin UASVM Animal Science and Biotechnologies*, 70 (1), 45-54.

Petrea S.M., Coada M.T., Cristea V., Dediu L., Turek Rahoveanu A.T., Zugravu A.G., Turek Rahoveanu M.M., Nocileta Mocuta D., 2016. A comparative cost-effectiveness analysis in diferent tested aquaponics systems. *Agriculture and Agricultural Science Procedia*, 10, 555-565.

Pillay T.V.R., Kutty, M.N., 2005. *Aquaculture, Principles and Practices*, 2nd Edition, Blackwell Publishing Ltd, Oxford, 630 p.

Pramanik M. H. R., Nagai M., Asao T., Matsui Y., 2000. Effects of temperature and photoperiod on phytotoxic root exudates of cucumber (*Cucumis sativus*) in hydroponic culture. *Journal of Chemical Ecology*, 26 (8), 1953-1967.

Quagrainie K.K., Valladao Flores R.M., Kim H., McClain V., 2017. Economic analysis of aquaponics and hydroponics production in the U.S. Midwest. *Journal of Applied Aquaculture*, 30 (1), 1-14, doi:10.1080/10454438.2017.1414009.

Quilleré I., 1994. L'intégration des cultures végétales dans les élevages piscicoles en eau recyclée. *Cahiers Agricultures*, 3, 301-8.

Quilleré I., Marie D., Roux L., Gosse F., Morot-Gaudry J.F., 1995. An artificial productive ecosystem based on a fish/bacteria/plant association. 2. Performance. *Agriculture, Ecosystems and Environment*. 53 (1), 19-30.

Ragnarsdottir K.V., Sverdrup H.U., Koca D., 2011. Challenging the planetary boundaries I: basic principles of an integrated model for phosphorous supply dynamics and global population size. *Applied Geochemistry,* 26, S303-S306.

Rakocy J.E., Masser M.P., Losordo T.M., 1992. Recirculating aquaculture tank production systems: integrating fish and plant culture, Southern Regional Aquaculture Center Pub. *SRAC Publication*, 454, 1-7.

Rakocy J.E., Shultz R.C., Bailey D.S., Thoman E.S., 2004. Aquaponic production of tilapia and basil: comparing a batch and staggered cropping system. *Acta Horticulturae*, 648, 63-69.

Rakocy J.E., Masser M.P., Losordo T.M., 2006. Recirculating aquaculture tank production systems: Aquaponics-Integrating fish and plant culture. Southern Regional Aquaculture Center. *SRAC Publication*, 454, 1-16.

Resh H.M., 2012. *Hydroponic food production: a definitive guidebook for the advanced home gardener and the commercial hydroponic grower*, Boca Raton (Florida), CRC Press, 560 p.

Renkui C., Dashu N., Jianguo W., 1995. *Rice-fish culture in China*, IDRC/CRDI, Ottawa, 276 p.

Réussir Fruits et légumes, 2019, https://www.reussir.fr/fruits-legumes/fraisereferences-le-hors-sol-fait-redecoller-la-fraise (accessed 11/02/2019).

Rey-Valette H., 2014. Some hints on the future of French aquaculture in 2040. *Cahiers Agricultures*, 23 (81), 34-46.

Rico Garcia E., Casanova Villareal V.E., Mercado-Luna A., 2009. Content of summer lettuce production using fish culture water, *Trends in Agriculture Economics*, 2 (1), 1-9.

Rojo F.G., Reynoso M.M., Ferez M., Chulze S.N., Torres A.M., 2007. Biological control by *Trichoderma* species of *Fusarium solani* causing peanut brown root rot under field conditions. *Crop Protection*, 26, 549-555.

Roosta H.R., 2011. Effects of foliar application of some macro and micronutrients on tomato plants in aquaponic and hydroponic systems. *Scienta Horticulturae*, 129 (3), 396-402.

Roosta H.R., 2014a. Effects of foliar spray of K on mint, radish, parsley and coriander plants in aquaponic system. *Journal of Plant Nutrition*, 37 (14), 396-402.

Roosta H.R., 2014b. Comparison of the vegetative growth, eco-physiological characteristics and mineral nutrient content of basil plants in different irrigation ratios of hydroponic: aquaponic solutions. *Journal of Plant Nutrition*, 37 (11), 1782-1803.

Roque d'Orbcastel E., Blancheton J.-P., Belaud A., 2009. Water quality and rainbow trout performance in a danish model farm recirculating system: comparison with a flow through system. *Aquacultural Engineering*, 40 (3), 135-143.

Rubio V.C., Sanchez F.J., Madrid J.A., 2005. Effects of salinity on food intake and macronutrient selection in European sea bass. *Physiology and Behavior*, 85, 333-339.

Salam M.A., Hashem S., Li F., 2014a. Nutrient recovery from in fish farming wastewater: an aquaponics system for plant and fish integration. *World Journal of Fish and Marine Sciences*, 6 (4), 55-360.

Salam M.A., Prodhan M. Y., Sayem S. M., Islam M. A., 2014b. Comparative growth performances of taro plant in aquaponics vs other systems. *International Journal of Innovation and Applied Studies*, 7 (3), 941-946.

Saufie S., Estim A., Tamin M., Harun A., Obong S., Mustafa S., 2015. Growth performance of tomato plant and genetically improved farmed tilapia in combined aquaponic systems. *Asian Journal of Agricultural Research*, 9 (3), 95-103.

Savidov N., 2004. *Evaluation and development of aquaponics production and product market capabilities in Alberta*. Ids Initiatives Fund Final Report, Crop Diversification Centre South, Alberta, 190 p.

Savidov N.A., 2005. Evaluation of aquaponics technology in Alberta, Canada. *Aquaponics Journal*, 37, http://www.aquaponics.com/wp-content/uploads/articles/Evaluation-of-Aquaponics-Technology-in-Alberta.pdf (accessed 5/03/2019).

Savidov N.A., Rakocy J.E., 2007. Fish and plant production in a recirculating aquaponic system: a new approach to sustainable agriculture in Canada. *Acta Horticulturae*, 742, 209-221.

Savidov N., 2014. Evolution of aquaponics design: from UVI to zerowaste commercial operation. Alberta Agriculture and Rural Development. For the aquaponics association, San Jose California, 131 p., http://www.aquaponicsiberia.com/wp-content/uploads/2016/10/nicksavidov-evolutionofaquaponics.pdf (accessed 28/02/2019).

Schmautz Z., Loeu F., Liebisch F., Graber A., Mathis A., Bulc T. G., 2016. Tomato productivity and quality in aquaponics: comparison of three hydroponic methods. *Water*, 8, 1-21, doi:10.3390/w8110533.

Schmautz Z., Graber A., Jaenicke S., Goesmann A., Junge R., Smits T.H.M., 2017. Microbial diversity in different compartments of an aquaponics system. *Archives of Microbiology*, 199, 613-620, doi:10.1007/s00203-016-1334-1.

Schneider O., Sereti V., Eding E.H., Verreth J.A.J., 2005. Analysis of nutrient flows in integrated intensive aquaculture systems, *Aquaculture Engineering*, 32, 379-401.

Seawright D.E., Stickney R.R., Walker R.B., 1998. Nutrient dynamics in integrated aquaculture-hydroponics systems. *Aquaculture*, 160, 215-237.

Selosse M.A., Baudoin E., Vandenkoornhuyse P., 2004. Symbiotic microorganisms, a key for ecological success and protection of plants. *Comptes Rendus Biologies*, 327 (7), 639-648 .

Sgherri C., Kadlecova Z., Pardossi A., Navari-Izzo F., Izzo E., 2008. Irrigation with diluted seawater improves the nutritional value of cherry tomatoes. *Journal of Agricultural and Food Chemistry*, 56, 3391-3397.

Sharma S.B., Sayyed R.Z., Mrugesh H.T., Thivakaran A.G., 2013. Phosphate solubilizing microbes: sustainable approach for managing phosphorus deficiency in agricultural soils. *SpringerPlus*, 2, 1-14.

Sheridan C., Depuydt P., de Ro M., Petit C., van Gysegem E., Delaere P., 2016. Microbial community dynamics and response to plant growth-promoting microorganisms in the rhizosphere of four common food crops grown in hydroponics. *Microbial Ecology*, 73, 378-393, doi:10.1007/s00248-016-0855-0.

Shete A.P., Verma A.K., Tandel R.S., Prakash C., Tiwari V.K., Hussai T., 2013. Optimization of water circulation period for the culture of goldfish with spinach in aquaponic system. *Journal of Agricultural Science*, 5 (4), 26-30.

Shete A.P., Verma A.K., Chadha N.K., Prakash C., Ahmad I., Nuwansi K.K.T., 2016. Optimization of hydraulic loading rate in aquaponic system with common carp (*Cyprinus carpio*) and mint (*Mentha arvensis*). *Journal of Agricultural Science*, 5 (4), 53-57.

Sikawa D.C., Yakupitiyage A., 2010. The hydroponic production of lettuce (*Lactuca sativa* L) by using hybrid catfish (*Clarias macrocephalus* × *Clarias gariepinus*) pond water: Potentials and contraints. *Agricultural Water Management,* 97, 1317-1325.

Simenoidou M., Paschos I., Gouva E., Kolygas M., Perdikaris C., 2012. Performance of a small-scale modular aquaponic system. *AACL Bioflux*, 5 (4), 182-188.

Singh D., Basu C., Meinhardt-Wollweber M., Roth B., 2014. LEDs for energy efficient greenhouse lighting. *Renewable and Sustainable Energy Reviews*, 49, 139-147.

Sirakov I., Lutz M., Graber A., Mathis A., Staykov Y., Smits T.H.M., Junge R., 2016. Potential for combined biocontrol activity against fungal fish and plant pathogens by bacterial isolates from a model aquaponic system. *Water*, 8 (518), 2-7.

Sirsat S.A., Neal J.A., 2013. Microbial profile of soil-free versus in-soil grown lettuce and intervention methodologies to combat pathogen surrogates and spoilage microorganisms on lettuce. *Foods*, 2, 488-498.

Slager, 2014. Modelling and evaluation of productivity and economic feasibility of a combined production of tomato and algae in Dutch greenhouses. *Biosystems engineering*, 122, 149-162.

Sneed K., Allen K., Ellis J.E., 1975. Fish farming and hydroponics. *Aquaculture and the fish farmer*, 1 (1), 11-18.

Somerville C., Cohen M., Pantanella E., Stankus A., Lovatelli A., 2014. Small-scale aquaponic food production: integrated fish and plants farming, FAO, Rome, 288 p., http://www.fao.org/3/a-i4021e/ (accessed 19/02/2019).

Specht K., Weith T., Swoboda K., Siebert R., 2016. Socially acceptable urban agriculture businesses. *Agronomy for Sustainable Development*, 36, 1-14.

Suhl J., Dannehl D., Kloas W., Banganz D., Jobs S., Scheibe G., Schmidt U., 2016. Advanced aquaponics: evaluation of intensive tomato production inaquaponics vs. conventional hydroponics. *Agricultural Water Management*, 178, 335-34.

Tamin M., Harun A., Estim A., Saufie S., Obong S., 2015. Consumer acceptance towards aquaponic products. *IOSR Journal of Business and Management*, 17, 49-64.

Tanguy H., 2008. Rapport final de la mission sur le développement de l'aquaculture, Ministère de l'agriculture et de la pêche, Ministère de l'écologie, de l'énergie, du développement durable et l'aménagement du territoire, 62 p.

Tavares J., Wang K-H., Hooks C.R.R., 2015. An evaluation of insectary plants for management of insect pests in a hydroponic cropping system. *Biological Control*, 91, 1-9.

Thorarinsdottir R., Kledal P.R., Skar S.L.G, Sustaeta F., Ragnarsdottir K.V., Mankasingh U., Pantanella E., Van de Ven R., Schultz C., 2015. *Aquaponics Guidelines*, 69 p., doi:10.13140/RG.2.1.4975.6880.

Timmons M.B., Ebeling J.M., 2007. *Recirculating Aquaculture*. Cayuga Aqua Ventures, LLC. 2nd edition, Ithaca, 948 p.

Todd J., 1980. Dreaming in my own backyard. *The journal of the New Alchemists*, 6, 108-111.

Tokunaga K., Tamaru C., Ako H., Leung P., 2015. Economics of small-scale commercial aquaponics in Hawaii. *Journal of the World Aquaculture Society*, 46 (1), 20-32.

Tran G., Heuzé V., Makkar H.P.S., 2015. Insects in fish diets, *Animal Frontiers*, 2, 37-44.

Trang N.T.D., Schierup H., Brix H., 2010. Leaf vegetables for use in integrated hydroponics and aquaculture systems: Effects of root flooding on growth, mineral composition and nutrient uptake. *African Journal of Biotechnology*, 9 (27), 4186-4196.

Trang N.T.D., Konnerup D., Brix H., 2017. Effects of recirculation rates on water quality and *Oreochromis niloticus* growth in aquaponic systems. *Aquacultural Engineering*, 78 (B), 95-104.

Treadwell D., Taber S., Tyson R., Simonne E., 2010. Foliar-applied micronutrients in aquaponics: a guide to use and sourcing. Horticultural Sciences Department, Florida Cooperative Extension Service, Institute of Food and Agricultural Sciences, University of Florida, HS1163, http://ufdc.ufl.edu/IR00003824/00001 (accessed 20/02/2019).

Trejo-Tellez L.I., Gomez-Merino F.C., 2012. Nutrient solutions for hydroponic systems, in: *Hydroponics, a standard methodology for plant biological researches* (Asao, T.), Intech, Rijeka, 254 p.

Truog E., 1947. Soil Reaction Influence on Availability of Plant Nutrients. *Soil Sciences Society of American Journal*, 11©, 305-308.

Turcios A.E., Papenbrock J., 2014. Sustainable treatment of aquaculture effluents - what can we learn from the past for the future? *Sustainability*, 6, 836-856.

Tyson R., Simonne J., Eric H.S., White J.M., Lamb M., 2004. Reconciling water quality parameters impacting nitrification in aquaponics: the pH levels. *Horticultural Society*, 117, 79-83.

Tyson R., Simonne J., Eric H.S., Treadwell D., 2008. Reconciling pH for ammonia biofiltration and cucumber yield in a recirculating aquaponics system with perlite biofilters. *HortScience*, 43 (3), 719-724.

Tyson R., Danyluk M.D., Simonne E.H., Treadwell D.D., 2012. Aquaponics, sustainable vegetable and fish co-production. *Proceedings of the Florida State Horticultural Society*, 125, 381-385.

Tyson R., Simonne J., Eric H.S., Treadwell D., 2011. Opportunities and challenges to sustainability in aquaponic systems. *HortTechnology*, 21 (1), 6-13.

Utkhede, R., 2006. Increased growth and yield of hydroponically grown greenhouse tomato plants inoculated with arbuscular mycorrhizal fungi and *Fusarium oxysporum f. sp. Radicis-lycopersici. BioControl*, 51, 393-400.

Valo M., 2013. Aquaculture production set to overtake beef. Article published in Le Monde on 16 December 2013, https://www.lemonde.fr/planete/article/2013/12/16/la-production-de-l-aquaculture-va-depasser-celle-du-b-uf_4334953_3244.html (accessed 28/03/2019).

Varmat A., 2008. Mycorrhiza, *state of the art, genetics and molecular biology, eco-function, biotechnology, eco-physiology, structure and systematics*, Springer, Berlin, 797 p.

Verdonck M., Taymans M., 2012. Sustainable food system. Employment potential in the Brussels-Capital Region. Final report of the research carried out for the Brussels Institute for Environmental Management. Centre for Brussels Regional Studies, 118 p.

Vergote N., Vermeulen J., 2012). Recirculation aquaculture system with tilapia in a hydroponic system with tomatoes. *Acta Horticulturae*, 927, 67-74.

Villarroel M. Alvarino J.M.R., Duran J.M., 2011. Aquaponics: integrating fish feeding rates and ion waste production for strawberry hydroponics. *Spanish Journal of Agricultural Research*, 9 (2), 537-545.

Weebly, The Green centre, [online],< http://www.thegreencenter.net> (accessed 25 January 2019).

Villarroel M., Junge R., Komives T., König B., Plaza I., Bittsanszky, Joly A., 2016. Survey of aquaponics in Europe. *Water*, 8, 468, doi:10.3390/w8100468.

Woensel L.V., Archer G., 2014. Ten technologies which could change our lives: potential impacts and policy implications. European Parliament, Scientific Foresight Unit (STOA), 28 p., http://www.europarl.europa.eu/EPRS/EPRS_IDAN_527417_ten_trends_to_change_your_life.pdf (accessed 28/02/2019).

Wongkiew S., Popp B.N., Kim H.J., Khanal S.K., 2017a. Fate of nitrogen in floating-raft aquaponic systems using natural abundance nitrogen isotopic compositions. *International Biodeterioration and Biodegradation*, 125, 24-32.

Wongkiew S., Hu Z., Kartik C., Woo Lee J., Khanal S.K., 2017b. Nitrogen transformations in aquaponic systems: A review. *Aquacultural Engineering*, 76, 9-19.

Xie K., Rosentrater K.A., 2015. Life cycle assessment (LCA) and Techno-economic analysis (TEA) of tilapia-basil aquaponics. Annual International Meeting, doi:10.13031/aim.20152188617.

Yildiz H.R., Robaina L., Pirhonen J., Mente E., Dominguez D., Parisi G., 2017. Fish welfare in aquaponic systems: its relation to water quality with an emphasis on feed and faeces - a review. *Water*, 9, 13, doi:10.3390/w9010013.

Zaidi A., Khan M.S., Ahemad M., Oves M., 2009. Plant growth promotion by phosphate solubilizing bacteria. *Acta Microbiologica et Immunologica Hungarica*, 56 (3), 263-284, doi:10.1556/AMicr.56.2009.3.6.

Zhang B., Tieman D., Jiao C., Xu Y., Chen K., Fei Z., Giovannoni J., Klee H., 2016. Chilling-induced tomato flavor loss is associated with altered volatile synthesis and transient changes in DNA methylation. *Proceedings of the National Academy of Sciences*. 113 (44), 12580-12585.

Zhang H., Li X., Yang Q., Sun L., Yang X., Zhou M., Deng R., Bi L., 2017. Plant growth, antibiotic uptake, and prevalence of antibiotic resistance in an endophytic system of pakchoi under antibiotic exposure. *International Journal of Environmental Research and Public Health*, 14 (11), 1336.

Zhang S-Y., Li G., Wu H-B.,Liu X-G., 2011. An integrated recirculating aquaculture system (RAS) for land-based fish farming: The effects on water quality and fish production, *Aquacultural Engineering*, 45, 93-102.

Zou Y., Zhen H., Zhang J., Xie H., Guimbaud C., Fang Y., 2015. Effects of pH on nitrogen transformations in media-based aquaponics. *Bioresource Technology*, 210, 81-87.

Zugravu G.A., Rahoveanu T.M.M., Rahoveanu T.A., Khalel S.M., Ibrahim R.A.M., 2016. The perception of aquaponics products in Romania. *International Conference Risk in Contemporary Economy*, "Dunarea de Jos" University of Galati, Faculty of Economics and Business Administration, 525-530.

Zweig R.D., 1986. An integrated fish culture hydroponic vegetable production system. *Aquaculture Magazine*, 12 (3), 34-40.

University reports

Baker A., 2010. Preliminary development and evaluation of an aquaponic system for the American Insular Pacific, Degree of Master of Science, Molecular biosciences and bioengineering, University of Hawaii, Manoa, 39 p.

Barrut B., 2011. Étude et optimisation du fonctionnement d'une colonne *airlift* à dépression - application à l'aquaculture, PhD thesis, speciality in energetics and process engineering, Université Montpellier II, Faculté des sciences et techniques du Languedoc, Montpellier, 154 p.

Danaher J.J., 2013. Phytoremediation of aquaculture effluent using integrated aquaculture production systems, Degree of Doctor of philosophy, Auburn University, Auburn, 203 p.

Goddek S., 2017. Opportunities and challenges of multi-loop aquaponic systems, Degree of Doctor, Biobased Chemistry and Technology, Wageningen University, Wageningen, 179 p.

Lapere P., 2010. A techno-economic feasibility study into aquaponics in South Africa, Degree of Master of Science, Engineering, University of Stellenbosch, Stellenbosch, 147 p.

McMurtry M.R., 1992. Integrated aquaculture-olericulture system as influenced by component ratio, Degree of Doctor of philosophy, North Carolina State University, Raleigh, 78 p.

Michaud L., 2007. Microbial communities of recirculating aquaculture facilities: interaction between heterotrophic and autotrophic bacteria and the system itself, PhD thesis, ecology of continental aquatic systems, Université Montpellier II, sciences et techniques du Languedoc, Montpellier, 142 p.

Pambrun V., 2005. Analyse et modélisation de la nitrification partielle et de la précipitation concomitante du phosphore dans un réacteur à alimentation séquencée, PhD thesis, process and environmental engineering, Institut national des sciences appliquées de Toulouse, Toulouse, 287 p.

Roque d'Orbcastel E., 2008. Optimisation de deux systèmes de production piscicole : biotransformation des nutriments et gestion des rejets, PhD thesis, sciences écologiques, vétérinaires, agronomiques et bioingénieries, Institut national polytechnique de Toulouse, Toulouse, 144 p.

Sanon A.A., 2005. Rôle des champignons mycorhiziens à arbuscules dans les mécanismes régissant la co-existence entre espèces végétales, diplôme d'études approfondies, sciences du sol, université Henri Poincaré, Nancy, 28 p.

Regulatory articles

Aquaculture compartment regulations

Arrêté du 1er avril 2008 fixant les prescriptions générales applicables aux installations, ouvrages, travaux ou activités soumis à déclaration en application des articles L. 214-1 à L. 214-6 du code de l'environnement et relever de la rubrique 3.2.7.0 de la nomenclature annexée au tableau de l'article R. 214-1

of the Environment Code (freshwater fish farms mentioned in Article L. 431-6) and repealing the Order of 14 June 2000, https://www.legifrance.gouv.fr/affichTexte.do?cidTexte=JORFTEXT000018663529 (accessed 28/02/2019)

Order of 1st April 2008 (OJ of 12/04/2008) defining the technical rules to be met by freshwater fish farms subject to authorisation under Book V of the Environment Code (ICPE).

Council Directive 2006 No 88/EC on animal health requirements for aquaculture animals and products thereof, https://eurlex.europa.eu/LexUriServ/LexUriServ.do?uri=OJ:L:2006:328:0014:0056:FR:PDF (accessed 28/02/2019).

Regulation (EU) No 304/2011 of the European Parliament and of the Council of 9 March 2011 amending Council Regulation (EC) No 708/2007 concerning use of alien and locally absent species in aquaculture, https://eurlex.europa.eu/LexUriServ/LexUriServ.do?uri=OJ:L:2011:088:0001:0004:FR:PDF (accessed 28/02/2019).

Commission Implementing Regulation (EU) No 1358/2014 of 18 December 2014 amending Regulation (EC) No 889/2008 laying down detailed rules for the implementation of Council Regulation (EC) No 834/2007, as regards the origin of animals used in organic aquaculture, aquaculture husbandry practices, feeding of animals used in organic aquaculture and products and substances authorised for use in organic aquaculture, https://eur-lex.europa.eu/legal-content/FR/TXT/PDF/?uri=CELEX:32014R1358&from=EN (accessed 28/02/2019).

Horticultural compartment regulations

Afssa, saisine n° 2009-SA-0288: avis de l'Agence française de sécurité sanitaire des aliments relatifs à l'évaluation des risques sur les effluents issus des établissements de transformation de sous-produits animaux de catégories 1, 2 ou 3 à des fins de réutilisation pour l'irrigation des cultures destinées à la consommation humaine ou animale, https://www.anses.fr/fr/system/files/EAUX2009sa0288.pdf (accessed 28/02/2019).

Arrêté du 2 août 2010, NOR: SASP1013629A, version consolidée au 18 février 2014 relatif à l'utilisation d'eaux issues du traitement d'épuration des eaux résiduaines pour l'irrigation de cultures ou d'espaces verts, https://www.legifrance.gouv.fr/affichTexte.do?cidTexte=JORFTEXT000022753522 (accessed 28/02/2019).

Avis de l'ANSES, réutilisation des eaux usées traitées pour l'irrigation des cultures, l'arrosage des espaces verts par aspersion et le lavage des voiries, ANSES, rapport d'expertise collective, 150 p., https://www.anses.fr/fr/system/files/EAUX2009sa0329Ra.pdf (accessed 28/02/2019).

General food hygiene and health regulations

Regulation (EC) No 178/2002 of the European Parliament and of the Council of 28 January 2002 laying down the general principles and requirements of food law, establishing the European Food Safety Authority and laying down procedures in matters of food safety, 24 p., https://eur-lex.europa.eu/legal-content/FR/TXT/PDF/?uri=CELEX:32002R0178&from=DE (accessed 28/02/2019).

Regulation (EC) No 852/2004 of the European Parliament and of the Council of 29 April 2004 on the hygiene of foodstuffs, 25 p., https://eurlex.europa.eu/LexUriServ/LexUriServ.do?uri=CONSLEG:2004R0852:20090420:FR:PDF (accessed 28/02/2019).

Commission Regulation (EC) No 2073/2005 of 15 November 2005 on microbiological criteria for foodstuffs, https://eur-lex.europa.eu/eli/reg/2005/2073/oj (accessed 28/02/2019).

Commission Regulation (EC) No 1881/2006 of 19 December 2006 setting maximum levels for certain contaminants in foodstuffs. https://eurlex.europa.eu/LexUriServ/LexUriServ.do?uri=OJ:L:2006:364:0005:0024:FR:PDF (accessed 28/02/2019).

Afssa, saisine n° 2007-SA-0174 : avis de l'Agence française de sécurité sanitaire des aliments concernant les références applicables aux denrées alimentaires en tant que critères indicateurs d'hygiène des procédés, https://www.anses.fr/fr/system/files/MIC2007sa0174.pdf (accessed 28/02/2019).

Authors

_____

Pierre Foucard, Former Research and Experimentation Engineer Aquaculture Department - Institut technique des filières avicole, cunicole et piscicole (ITAVI)

Aurélien Tocqueville, Aquaculture Department Manager - Institut technique des filières avicole, cunicole et piscicole (ITAVI) tocqueville@itavi.asso.fr

Co-authors/Participants

_____

Jean-François Baroiller, Former Researcher, UMR116 ISEM, Aquaculture - Centre de coopération internationale en recherche agronomique pour le développement (CIRAD)

Matthieu Gaumé, Former Research and Experimentation Engineer, Aquaculture Department - Institut technique des filières avicole, cunicole et piscicole (ITAVI)

Laurent Labbé, Former Research Engineer and Director of the INRA Monts d'Arrée Experimental Fish Farm - Institut national de la recherche agronomique (INRA)

Catherine Lejolivet, aquaculture teacher and "Water, aquaculture, aquaponics" project manager - Établissement public local d'enseignement agricole (EPLEFPA) Lozère - La Canourgue site Catherine. Lejolivet@educagri.fr

The authors would like to thank:

– Bernard Darfeuille, former technical manager of the RATHO/ASTREDHOR station;
– Serge Lepage, former RATHO station manager
– and the entire RATHO team
– Anne Richard, Former Director of ITAVI;
– students on work experience in the aquaculture department of ITAVI, working in aquaponics;
– DGER/CASDAR, Ministry of Agriculture DPMA (FEAMP), France Agrimer, for support for the APIVA® and Aquaponie projects led by ITAVI;
– all the project leaders and people interested in and passionate about aquaponics whom we have met in the course of our work.

Index

www.ingramcontent.com/pod-product-compliance
Lightning Source LLC
Chambersburg PA
CBHW040136200326
41458CB00025B/6285